WHO KILLED NEW ORLEANS?

WHO KILLED NEW ORLEANS?

✦

Mother Nature vs. Human Nature

Diane Holloway, Ph.D.
With Johannes Spreen and Bob Cheney

iUniverse, Inc.
New York Lincoln Shanghai

WHO KILLED NEW ORLEANS?
Mother Nature vs. Human Nature

iUniverse books may be ordered through booksellers or by contacting:

iUniverse
2021 Pine Lake Road, Suite 100
Lincoln, NE 68512
www.iuniverse.com
1-800-Authors (1-800-288-4677)

ISBN-13: 978-0-595-37391-8 (pbk)
ISBN-13: 978-0-595-81786-3 (ebk)
ISBN-10: 0-595-37391-7 (pbk)
ISBN-10: 0-595-81786-6 (ebk)

Printed in the United States of America

Contents

PREFACE ix

DEDICATION xi

FOREWORD xiii

CHAPTER 1   Why Are We Blaming Everyone for Killing New Orleans? 1

CHAPTER 2   Who Can Control "Ole Man?" 20

CHAPTER 3   Facts That Were Denied 33

CHAPTER 4   Police Had Two Roles 50

CHAPTER 5   Politicians Had Trouble Leading 66

CHAPTER 6   Press Probably Saved Lives 80

CHAPTER 7   Pressure Groups Created Pre-disaster Problems 88

CHAPTER 8   People Are Only Human 93

CHAPTER 9   Are We Stupid to Rebuild New Orleans? 114

CHAPTER 10   What Can We Learn from the Past? 125

AFTERWORD By Johannes Spreen 141

AFTERWORD By Bob Cheney 147

RESOURCES 149

APPENDIX A   Community Disaster Assessment 153

APPENDIX B   Personal Disaster Assessment 159

About the Author 171

Index 173

ACKNOWLEDGEMENTS

I want to thank so many people who dropped everything and helped in the production of this book so quickly after Hurricanes Katrina and Rita devastated many areas of the Gulf Coast and New Orleans.

I appreciate Ken Jacuzzi for offering his very thoughtful analysis of events and their consequences in the Preface. He is the author of *Jacuzzi: A Father's Invention to Ease a Son's Pain,* published 2005.

I am very indebted to Chief Robert Biscoe of the Fire District of Sun City West, who helped me with information and wording about emergency services. I am also grateful to Deputy Chief Steve Penney for directing me to Captain Scott Sherck and Tait Mitton who described their four weeks in Mississippi working for FEMA. They aided volunteer fire department in distributing food, water, ice and information and helped register over 20,000 families for FEMA-assisted accommodations and other services.

I want to thank Dr. Don Farrior, minister and supervisor of chaplains at Del Webb Memorial Hospital for helping make clear the role of humor in calamities and stressful circumstances. He has just written his own book about humor in the church and chaplaincy called *I Did Not Burn the Church Down...I Only Started the Fire,* published in 2005.

Most of all I want to thank my co-authors who contributed so much help in the production, the content, and the information included in this book. Johannes F. Spreen, former police commissioner of Detroit, former Oakland County sheriff and professor, was able to offer so very much to the law enforcement perspective of events in disasters. Historian Bob Cheney was able to look at the long view and the context into which this disaster was placed in American and world history. Additionally, he scrutinized the text carefully and made a multitude of needed changes and improvements.

I would also like to thank the production staff at IUniverse Publications, starting with Rifka Keilson and Ron Amack for their help in putting this book together so quickly for us.

I thank you all and hope that this little contribution to the most expensive catastrophe in American history adds some new thoughts about how we might prevent such devastation and loss of life in the future.

PREFACE

As we are all painfully aware, New Orleans, Mobile, Port Arthur, and many other communities along the Gulf Coast have been unimaginably devastated by Hurricanes Katrina and Rita. Katrina left hundreds dead and hosts of people without homes or jobs to return to.

The images and stories I have seen unfold in the media have drained me, leaving me with a sense of impotence, but also a glimmer of hope. Certainly, as was the case in the post-9/11 tragedy, men and women of every walk of life have come forward and unselfishly reached out to help their fellow man. There can be no doubt that an American's basic nature is one of goodness and giving, especially during times of mortifying trial that put in jeopardy our very survival. My personal hope, though, stems not just from the good I see in us, but from the lessons we can and should learn from this tragic, monumental catastrophe of nature.

First, it shows unequivocally how very precious every single day in one's life is.

Second, we must permanently prioritize the care of the natural environment in which we all live, and refuse to let precious, protective wetlands and other gifts from our earth be destroyed and replaced by reckless, thoughtless development.

Third, we owe it to all our citizens to once and for all establish a system of universal healthcare that will cover every single citizen and legal resident of the United States. Only such a system can properly protect us in times of rapidly spreading infections, diseases, and biological disasters.

Fourth, we need to seriously consider establishing the requirement that all able-bodied Americans perform between 12 and 18 months of

National Service for their country. It will be up to each individual to choose whether his or her National Service be part of or outside the U.S. military. National Service participation—whether in our parks or forests, schools, hospitals, or building infrastructure—will ensure that this nation will always have a sufficient number of prepared personnel to assist in catastrophic emergencies, whether caused by nature or man-made.

Fifth, and perhaps most importantly, we must change and re-create our current leadership culture, including and starting from the President of the United States, captains of industry, and leaders of local government, law enforcement, healthcare, education, and the media. We are, unfortunately, inundated with self-serving behaviors, from those who loot and shoot to those in positions of power who cast aspersions and dodge blame.

It becomes imperative, instead, that our fifth lesson should teach us that we must extol and practice—commencing with our leaders—a culture which makes preeminent brotherhood, mutual respect and caring for one another as well as the planet we are blessed to live on. I urge us all, believers or not, to pray for and/or help everyone who has experienced suffering from this disaster. I hope that all of us individually and collectively will finally comprehend the reason why our Native American Indian brothers and sisters have always called this planet in which we live, Mother Earth. That is recovery, better future survival, and life as I see it.

Kenneth Jacuzzi, Paradise Valley, Arizona, author of *Jacuzzi: A Father's Invention to Ease a Son's Pain,* published 2005.

DEDICATION

We would like to dedicate this book to the emergency service personnel who risked their lives to save victims during Hurricanes Katrina and Rita. These people maintain their training, practice their skills, and stand ready to save us in spite of ourselves. Usually they are paid professionals but consist also of volunteer fire departments and others. They all faced terrible odds as they fought raging flood waters, incredible winds, emergency vehicle and aircraft accidents that claimed lives, angry and armed victims, and sometimes just a lack of gratitude by survivors. We thank you for your selfless service.

Dr. Diane Holloway
Johannes F. Spreen
Bob W. Cheney

FOREWORD

Former Detroit Police Commissioner Johannes Spreen called me the second morning of Hurricane Katrina saying, "You've got to do a book on this terrible disaster. There are so many things happening that require some psychological understanding and you can do that. I'll help in any way I can." He knew of my psychology training and work with individuals, governments, and corporations.

He had been glued to the news coverage just as I had. This was a human tragedy just as compelling as the attack of the Twin Towers, Pentagon and the crashed plane of 9/11/01. News coverage was unrelenting and images of flooded Gulf Coast scenes kept coming day and night.

The worst flooding of all was in New Orleans. I, like almost everyone who has been there, have great memories of that old city. My husband had his first honeymoon there, not with me, but we went back several times during our marriage.

The French Quarter was our favorite place. We couldn't forget how the waiter in the Court of Two Sisters tossed the dregs of my coffee in the beautiful wishing well fountain in the center of the restaurant to refill my cup. We recalled a horse-drawn carriage ride by an impossible to understand Cajun. Our mornings would have been incomplete without café au lait and beignets from Café du Monde.

Even though our hotel room (where Tennessee Williams supposedly wrote *Streetcar Named Desire*) wasn't completely clean, we kidded that old Tennessee appeared to have left a few hairs in the bathroom shower stall. Who can forget the cuisine in that old French, Spanish and American city of gluttony? The blues and jazz draw so many in the evenings that people line the sidewalks to listen when they can't squeeze into night spots.

And almost everyone who has visited old New Orleans has had a Hurricane at Pat O'Brien's. That was just the name of a drink to most of us, until Katrina gave it a whole new meaning.

Our memories are just our own. Others would tell you about Mardi Gras and open sexuality and others would tell you about the Mississippi River and still others would describe those things that only residents and natives could know—the good, the bad, and the ugly.

So I decided to tackle the disaster and look at what we could learn from it. People already began to learn as Hurricane Rita quickly followed Hurricane Katrina, and undoubtedly countless lives were saved because of lessons learned. While we can't prevent Mother Nature and disasters, perhaps we can improve our human nature and our methods of saving ourselves.

Dr. Diane Holloway

1

Why Are We Blaming Everyone for Killing New Orleans?

"The purpose of defensive mechanisms is to avert dangers…Each person merely selects certain ones which are repeated through life whenever a situation evokes them…There really is a resistance to discovering the truth about ourselves."

—*Sigmund Freud*

Hurricane Katrina, quickly followed by Hurricane Rita, is the story of human nature against Mother Nature. We have nothing better than our own human nature to battle the elements and to battle each other. Being only human, when things go wrong, we use various coping methods that have become part of human nature in virtually every person on earth. Some of those coping methods are better than others. These hurricanes showed both our good and bad coping mechanisms.

Katrina and Rita exposed the lack of preparation and the ineptitude of some rescue efforts along the Gulf Coast but especially in New Orleans. We began to blame people and agencies. When I say "we," I mean all of us, from the President, on down to Federal Emergency Management Agency (FEMA) leaders, governors, mayors, policemen, reporters, politicians, rescuers, victims, and the man in the street. It's just human nature. We can dodge blame and responsibility ourselves if only we can find someone else to blame it on.

The Blame Game

We also tend to lie about our role in creating a disaster or our lack of preparation for it. Yes, we can say we did something or we didn't do something crucial to protect people, but if we were hooked up to a lie detector, it would be going crazy because our conscience knows the real truths.

Sometimes we don't tell an outright lie, we just rationalize things and make it sound like everything was done in a very reasonable way. Inside, we realize that a plan wasn't really well thought out. Someone, probably a dentist, came up with a phrase for this—"Truth decay!"

The last thing we ever want to admit is that we denied facts. That makes us sound so dumb. The city leaders, the protectors who make plans for disasters, and the politicians who control where money is spent never want to admit that they denied the fact that disasters happen. But we use denial all the time.

Think you don't use denial? Consider these things. We don't make out our wills because we'll *never* die. We don't mind using up all the forests to make products because trees will *always* be plentiful. We don't mind if fishing industries have exhausted 90% of certain fish species because fish will *always* be there. We don't tell people we love them very often because there will *always* be time. Young soldiers don't mind going to war because they'll *never* die. We don't mind turning down the funds to protect New Orleans residents from floods or making plans for getting them to safety because floods will *never* happen in our term of office. Yes, we deny reality, deny facts, and we deny what happened in the past.

We try to play the Blame Game with Katrina, but there is no *enemy* to focus on like the terrorist attacks of 9/11/01. People can handle calamities better when they can focus their anger and their efforts on a foe. Hitler gave his people someone to blame for poverty—the Jews. Osama bin Laden gave his Muslims someone to blame for their poverty—the heretics or non-Muslims. Bad weather is a poor substitute for an opponent. We want to get mad at a person or persons like Jews or Muslims or Indians or Blacks or Adolf Hitler or Osama bin Laden or George W. Bush.

First, let us look at some of the people who have received blame for problems resulting from Hurricane Katrina. It is almost like fairy tales where if one can discover the name of the secret villain, the evil power of the villain to do harm is broken. So in this most expensive disaster in American history, names have been named like never before.

Here is what Democratic National Committee Chairman Howard Dean said:

> While everyone along the Gulf Coast faced unimaginable hardships in the aftermath of Hurricane Katrina, the Bush Administration's failed leadership on homeland security and emergency preparedness took an especially heavy toll on Americans with disabilities, particularly those with low incomes. From an evacuation and disaster relief effort that failed to address their needs to the challenges of finding accessible housing, health care and prescription drug coverage in the weeks since the storm, the unique chal-

lenges confronting Katrina survivors with disabilities have been staggering, and the Bush Administration's response has been unacceptable.

The hallmark of this great nation is our commitment to ensuring that the most vulnerable among us are not left behind in times of need. But the harsh reality is that too many Americans with disabilities fell victim to the Bush Administration's failed leadership and poor planning. Four years after so many Americans with disabilities were unable to escape the World Trade Center, this tragedy demonstrates the Bush Administration's failure to learn the lessons of September 11th as they relate to the disabled.

We have an opportunity to address the problems revealed in the aftermath of this tragedy. Americans need real leadership that includes a reconstruction effort that includes the needs of Americans with disabilities, one that provides a model for a system in which Americans with disabilities are integrated into their homes and communities and not forced into nursing homes and institutions. President Bush and Republicans in Washington should join Democrats in working to ensure that Americans with disabilities are fully integrated into our society and included in our emergency preparedness plans, so that, moving forward, they are never again left behind.

Newspapers report that many, including Dean, want to blame President Bush, whom they say was too slow to respond. They have already blamed him with having a "blind spot" about the war in Iraq, as if he is a reasonable man and only unable to recognize facts in that one area. But when more blame was being laid on his shoulders for the handling of Katrina, he surprised everyone by accepting blame and admitting a mistake. Perhaps Bush was following President John Kennedy's acceptance of blame for the ill-planned Bay of Pigs invasion of Cuba. That acceptance won Kennedy praise for his honesty.

On September 13, 2005, in a brief news conference, President Bush said,

> Katrina exposed serious problems in our response capability at all levels of government, and to the extent that the federal government didn't fully do its job right, I take responsibility...I want to know how to better cooperate with state and local government to be able to answer [whether we are] capable of dealing with a severe attack or another severe storm. And that's a very important question. And it's in our national interest that we find out exactly what went on so that we can better respond.

In the 2000 presidential election debates, Bush chided Al Gore saying that natural disasters were "a time to test your mettle." Many began questioning whether the Bush administration's mettle had measured up to their promises.

Early in his presidency, President Bush downgraded the Federal Emergency Management Agency (FEMA), the entity designated to cope with national emergencies, by wrapping it under the new Homeland Security Department. Such agencies, the reasoning apparently went, were looking for government handouts. That was the very year that FEMA warned that a hurricane hitting New Orleans was one of the three "likeliest, most catastrophic disasters facing this country."

Speaking of FEMA, there was so much blame placed on Michael Brown, FEMA director, that he was removed from directing activities in New Orleans. Within two weeks after the hurricane, Brown resigned and was quickly replaced.

President Bush appeared to want to deflect accusations of racism in handling the Katrina disaster early on. Although Condoleezza Rice is Secretary of State, which has more to do with foreign policy than domestic policy, she announced that racism had nothing to do with the Katrina response. She then received criticism for being used by the White House to defend accusations that the Bush administration is racist.

Many other countries have done their share of criticizing during and after the hurricanes hit. They imply that this predictable natural calamity has exposed America's shortcomings. America, some countries say, sees itself as a nation of uniquely hardy people predestined to be omnipotent against the forces of nature or Saddam Hussein-style dictators. Americans believe that hardship can be overcome by hard work which will be rewarded. Yet Katrina and Rita showed the fragility of the U.S. disaster preparation and the need for leadership to replace the individualistic Protestant work ethic.

The sense of American omnipotence and the pride that American mastery can overcome all obstacles and evils presented by man or by God prevails, say the newspapers of some nations. They claim it shows up in the plan to rebuild New Orleans in exactly that same spot, which demonstrates a refusal to learn the lessons of history or understand the forces of nature. They caution that New Orleans was a disaster waiting to happen just as, one day, a monumental earthquake will hit Los Angeles or San Francisco. In the days immediately following Katrina and Rita, questions about whether to rebuild New Orleans in the same place seemed defeatist and curiously un-American. Some foreign leaders suggest that these lessons of Katrina and Rita are humbling but may cause Americans to strengthen life at home rather than trying to straighten out the rest of the world.

Hurricanes Katrina and Rita Exposed Real Problems

Federal government officials have done their share of blaming. They blamed both state and local officials, and those officials, chafing at the bit, have in turn blamed the federal government.

We all remember that after 9/11, there was a surge in patriotic feeling because everyone felt they were in the same boat. Hurricane Katrina exposed the fact that in New Orleans, everyone wasn't in the same boat. The poor and mainly black people died or suffered and are still suffering, whereas the rich and mainly white people reached safety. Thus, many have blamed those who created and maintained the great divide between the rich whites and poor blacks in New Orleans.

Katrina exposed yet another problem, also identified in 9/11. That problem is about the different parts of government that are supposed to work together to be effective but have difficulty doing that. So we tend to blame the various government agencies.

Four years ago, the September 11th attacks and Bush's leadership helped him win a second term. The attacks by four planes on the Twin Towers and Pentagon, which killed nearly 3,000 people, united political parties. Those parties resolved to protect the country better, and created the Homeland Security Department. That agency was to deal with both natural disasters *and* terrorist attacks. They approved billions of dollars and agreed on major anti-terrorism legislation. However, political parties began to split again after the invasion of Iraq and after probes revealed that perhaps federal agencies could have done more to prevent the terrorist attacks.

The responsibility for responding to natural disasters is shared by many agencies at all levels of government, many of whom were blamed in the aftermath of Katrina. The first reaction to emergencies comes from local government. It plans and arranges for evacuation, shelter, and first response by police, fire, and medical personnel. Flood-protection levees (in Louisiana) are the primary responsibility of local levee boards.

Two days before Katrina, local leaders like New Orleans Mayor Ray Nagin urged citizens to board up their homes, fill their gas tanks, and gather their medications. For those without cars or transportation, city leaders designated the Superdome, convention center and other sites as temporary shelters and arranged for bus pickups throughout the city.

At first, Nagin did not make the evacuation mandatory, and he has been blamed for that oversight. Announcements about what should be done with those in nursing homes or the homebound were sketchy. Apparently, well over 130,000 residents had no transportation, and the city didn't have sufficient help for them to get to safety. By the next day, the mayor made evacuation mandatory, saying that the storm would probably break the levees.

Local first responders came in for their share of criticism as well. Problems began to develop when many police officers, some busy relocating their own families, responded unpredictably. Some did not show up for work where expected, some (according to a few eyewitnesses) joined in the looting to help citizens get what they needed, and some reportedly looted for their own selfish purposes.

The local citizens played their role in creating problems. Those who refused to leave their homes were blamed for creating the need to rescue them. Those who began looting almost immediately were blamed more than those who looted later to satisfy their need for resources to survive. Within days, gunshots were fired at some rescue workers. Police were ordered to protect rescue workers, to arrest looters, and to shoot threatening armed citizens. After the first few days, however, criticism of the victims has been minimal.

Traditionally when local resources are overtaxed, state government is mobilized to assist with strategies and manpower. Initially, some 3,700 National Guard troops, under the control of the governor, were mobilized. Their numbers had, of course, been depleted by deployments to the Iraq war. So the government was blamed (again) for the Iraq war. State government earlier had been a source of matching funds that, combined with federal money, went toward building levees and legislators were blamed for inadequate levee funding.

Gov. Kathleen Blanco declared a state of emergency on August 26th as the storm gathered strength. She had already contacted President Bush for help because the crisis was more than state and local government could handle. She also warned citizens to evacuate, and ordered the Transportation Department and police to direct traffic flow. Once Hurricane Katrina hit land on August 28th, local police and state troops were criticized for being slow to get to impacted areas to restore order and stop looting. More troops were summoned from other states.

Some have criticized the governor and the federal agencies saying there was a delay because of disputes between the governor's office and

Washington about who had jurisdiction—who would command the Guard troops.

According to inter-governmental agreements, when disasters overwhelm state and local resources, FEMA takes over the response, providing supplies and helping with cleanup and aid to those whose homes are destroyed. FEMA can call upon other federal agencies such as the Defense Department for troops, air and sealift help. FEMA was not only blamed for being slow to assist, but for wanting to have ultimate coordination of rescue efforts. However, if some group does not take charge of coordinating activities, people and agencies will work at cross purposes.

The storms also revealed gaps in the National Response Plan which was supposed, after 9/11, to provide swift delivery of federal support in response to catastrophic incidents. The Plan was published in January, 2005, by the Department of Homeland Security and FEMA.

Sen. Hillary Rodham Clinton of New York questioned whether including FEMA under the Department of Homeland Security during the anti-terrorism reorganization had weakened its disaster-relief role.

The U.S. Army Corps of Engineers sent its personnel to safer ground as Katrina hurricane neared. They were blamed when it was found that no one was checking the levees to assess their strength, and decisions were delayed lacking that crucial information.

Two days *after* the hurricane hit, Bush ended his vacation a couple of days early and returned to Washington. He was blamed for, among other things, remaining on vacation as long as he did. He told ABC News on September 1st, "I don't think anybody anticipated the breach of the levees."

In a somewhat defensive posture to avoid blame, the White House issued a fact sheet to reporters describing their actions, which included sending in 50 medical assistance teams, 25 search-and-rescue task forces, and 1,700 trucks to transport supplies. Eight Navy ships were headed to the Gulf of Mexico, but could not arrive for several days.

The American Health System Was Lacking

The American health system has come in for much criticism since Hurricane Katrina. Health services have shown inadequate preparation to handle a large-scale public-health emergency. Health officials were accused of focusing on bio-terrorism and ignoring the nature of America's decaying public health system. This decay was addressed in a report issued in 2002 by the Institute of Medicine.

The report claimed that serious under-funding and political neglect had hurt governmental public health agencies. It criticized laws and regulations governing public health that had fragmented health responsibility among officials at all levels of government. The report stated that the uneven distribution of resources would render the health system unable to manage a large-scale emergency and that prediction proved true in Hurricane Katrina.

This is likely to cause Congressional leaders to expand federal funds for the Medicaid health program for the poor, both to Gulf Coast states as well as to states that received evacuees. Currently, states share in the cost of Medicaid with the federal government, but as some have pointed out, the states affected by Katrina will be unable to handle their matching funds due to the financial crisis caused by the storm.

At New Orleans Charity Hospital, hundreds of patients and medical staff were trapped for more than four days with accumulated sewage and contaminated water, no electrical power, clean water or medical supplies. Meanwhile, another hospital (Tulane University Medical Center) hired helicopters to evacuate its patients and family members.

Dead bodies were found several days later in still another hospital and a nursing home. Investigations are still ongoing but first reports indicate that elderly patients could not survive sweltering heat and had no help from electronic medical devices when power was lost. Overall, some 13 nursing homes and 6 hospitals were investigated for reports of euthanasia but to date no such action has been found.

When the water was finally pumped away, the number of corpses was not as high as was estimated. Just as 9/11 deaths were overestimated, so were Katrina's, fortunately. In fact, the total death toll throughout the Gulf Coast has not yet reached 1,300. Hurricane Rita had minimal deaths yet discovered from flooding. Nevertheless, the breakdown of the public-health infrastructure and civilized society was clear. Ambulance and repair crews were shot at, looting occurred, and some elderly and infirm were left to fend for themselves.

Emergency response trauma teams from other states arrived to the surprise of federal disaster officials in New Orleans who were unprepared to dispatch them effectively. The problem of self-dispatched emergency workers had compounded difficulties for personnel at the Twin Towers in New York on 9/11. People wanted to help and came from across the country and across the world, arriving in overwhelming numbers there.

Who Wanted To Help Us?

The victims of the hurricane were overwhelmingly black and poor. This reflected the population of New Orleans (over 70% black) and the public health and emergency systems that failed them, both before and during the crisis. Immediately after the storm, there was no 9-1-1. There was a non-availability of health care when infectious disease outbreaks caused by the decomposing corpses in flood waters worried public health experts. Blame was laid at many doorsteps for those occurrences.

At first, many Americans were heard to say that although we help other countries when they have crises, tsunamis and earthquakes, nobody ever reaches out to help us. The United States, within a week after the hurricane, announced that it would welcome help from other nations.

The offers of help had already started to pour in, and included many long-term allies (Canada, Great Britain, France, Germany, Australia, Japan) but also some surprising other sources. Venezuela and Cuba offered help including doctors, relief workers, and medicines. Fidel Castro claimed to be quite hurt that Cuban help was not accepted.

Large entities such as NATO, OAS, WHO, the Vatican, and other groups offered help.

Some specified where they would offer help such as Australia's money was to go to the American Red Cross. Canada sent vessels loaded with relief supplies and personnel. China sent $5 million for victims, Iran offered "humanitarian aid," and Israel sent health and defense officials to help coordinate aid. Mexico sent truckloads of water, food, medical supplies, ships, helicopters, and amphibious vehicles. The Netherlands sent teams for inspecting dikes. Saudi Aramco donated $6.5 million to the American Red Cross to support relief efforts. The list of countries offering help continued to grow during the first weeks of the devastation. These donations did not stop many from continuing to criticize other countries for not doing more.

When the Going Gets Tough, Do the Tough Get Going?

The Blame Game was the first reaction of so many people as they rushed to evade responsibility and criticism by pointing fingers elsewhere. But blame is only one of several defense reactions that people use when they feel under pressure. The more important and more deadly defense mechanism of denial of facts has yet to be explored. We hope that when the going gets tough, the tough get going, but sometimes they get going the wrong way.

In my 2002 book, *Analyzing Leaders, Presidents and Terrorists,* I described some of the healthy and unhealthy defense mechanisms or ways of handling stress used by Harry Truman, Adolf Hitler, Mao Tse-tung, Nikita Khrushchev, Fidel Castro, John Kennedy, Lyndon Johnson, Golda Meir, Richard Nixon, Ronald Reagan, Margaret Thatcher, Saddam Hussein and Osama bin Laden. Everyone uses techniques to cope with problems. That's our human nature.

We would like to hope that our leaders use very effective coping techniques since we assume they may have more problems and more stress than the average person. But who knows how our leaders will handle stress until it comes, and then their coping methods may be visible to all. That's their human nature.

Stress comes in many forms. There can be a general hyperactivity and quickness to react to anxiety or any sign of danger. Such people may be called hyperactive and may not have reliable methods of reducing stress. Some people, for example, tend to hold in all anger, fearing rejection by others if they show anger. They may passively show their anger by continually being late or obstructing plans or even cutting their wrist to retaliate against someone.

Others may develop physical symptoms or imagined illnesses such as hypochondriacs, who may then elicit sympathy from others for their "condition."

Others may become overly anxious when given more responsibility than they feel capable of handling. They may deliberately err to remove the mantle of responsibility from their shoulders.

Some people feel anxious when others don't seem to accept them or their orders, which throws their self-esteem into doubt. Reporters covered a meeting of 400 firefighters (held in Atlanta) sent to help rescue people. (A FEMA request for 1,000 two-man teams from U.S. fire departments had been sent out.) They reported the objections of the firefighters to being lectured by FEMA officials on equal opportunity, sexual harassment, and customer service. One said, according to reporter Mark Thompson writing in *Time* magazine, "This is ridiculous. Our fire departments and mayors sent us down here to save people, and you've got us doing this?" (39) The defensive FEMA leader tried to restore order by getting on a chair and telling the firefighters that they were now employees of FEMA and they were to follow orders.

How do leaders handle problems, anxieties and stresses, which are inevitable in their chosen role? Stress upsets one's equilibrium. Anxiety rises in anticipation of danger. Sometimes, an unrealistic perception of events causes more fear than is appropriate to the situation. Did Mayor

Ray Nagin overreact when he ordered 25,000 body bags, some 25 times the amount needed?

Some magazines reported that when President Bush is told that he is wrong by an aide, he becomes angry. If a leader surrounds himself by people who reassure him that all is well when it may not be, he is removed from critical information and is more likely to make wrong decisions. Such a reaction, if true, suggests that Bush's method of handling bad news is like the king who killed the messenger because he didn't like the message. Was Bush's frustration during the Katrina disaster a reaction to growing criticism of his presidency?

Unless a coping mechanism reduces anxiety and stress, it doesn't really calm the person. As stress and anxiety continue and frustration builds, some people retreat to coping methods which are inappropriate for normal adults. These methods don't remove the stress or help the person regain their equilibrium. In fact, they often make things worse.

The methods used to handle stress become habitual. The patterns used most often are what we commonly call "character." Parents often reward children for mature coping habits that remove problems and improve functioning. They might reward good grades when a child substitutes scholarly achievement for lack of sports ability. They often punish children who use inadequate or childish coping mechanisms, which might include temper tantrums, making excuses, risky sexual behavior, or drug use.

Some people may use fantasies or dreams to create pleasing scenarios or situations or even imaginary companions. This may be seen in those who fear rejection if they associate with real people. If they are rejected, self-doubt may cause them to feel inadequate or even worthless.

Some people project their own faults onto others, collect injustices, expect enemies, and become paranoid as they try to protect themselves from stress. This might explain some ne'er-do-wells who grabbed guns and shot at rescue workers during the height of the Katrina flood. Their character or coping patterns were visible for all to see and to criticize.

Of course, people would rather handle stress in laudable ways. What are some of those? The most useful coping techniques would involve serving others (altruism) and this was seen by many who selflessly helped others during the Gulf Coast disaster, even at the risk of their own safety.

Other useful coping techniques are anticipation and planning for problems. This was seen in many who reviewed hurricane disaster plans early on and set up written protocols, and training exercises. It was also seen by those leaders who quickly located disaster protocols and went by them, as much as they were able. This was done by Louisiana Governor Blanco, but when she got to the step of calling for federal help, protocols did not describe how to operate when the President could not be reached. Also, she learned later that she might have been more successful if she had requested specific help and numbers, since she was closer to assessing the needs than the White House.

Other successful coping techniques involve asceticism or pride in avoiding temptations. Those who broke into stores and took items such as televisions and guns yielded to temptation for personal gain and power. Those who passed up opportunities to take things once store windows were broken demonstrated asceticism or the ability to resist temptation.

Another coping mechanism is humor and making sport of problems. After Katrina hit Biloxi, Mississippi, Gary and Valentina Stillwell sat under a tarp in what was once the garden of a home built in 1855. All that was left were three front steps. The house they were attached to floated away. Gary told reporters that they rode their house for two and one half blocks. As they sat on the roof and looked around, they saw a little outcropping of a roof, shingles, and a bunch of trees that finally stopped their progress. The reporter asked what it was like to be in your house as it was moving down the street. Valentina said they kept watching everybody else's house float by and said things like, "Isn't that roof Ann and Art's, or Ted and Sandy's?"

Then Gary added, with a perfectly timed light touch, that they had been thinking about moving before the storm—but not exactly in this manner.

Another good example of humor after Katrina was shown by Greg and Crissy Avery in Covington, Louisiana. The Averys were sheltered in their log cabin during the storm, and Katrina blew a five-foot section of cedar down through their car, spearing the windshield before lodging in the dashboard. The tree left room for Crissy to get in the car and drive it about. The sight of the speared car with about three feet of tree trunk sticking out of the windshield caused many laughs and much humor in Covington during this incredible tragedy. Crissy said more people asked how it happened than about whether she was okay, but she knew the importance of a little diversion from problems.

The ability to laugh at oneself even in the worst of crises was demonstrated by President Ronald Reagan after he was shot in the chest very near the heart. As the doctors prepared to operate on him, he said with a twinkle of a smile, "I hope you're Republicans." Perhaps he was comforted when his doctors responded, "Today, Mr. President, we're all Republicans."

Are We All in the Same Boat?

At first, some events are too horrible to laugh about and actions must be quick and correct. There may be no time to lighten the burden with a word of humor, but even in the thick of battle or the morbidity of death camps, humor helps people get through ordeals. It returns a sense of companionship and "we're all in the same boat" to a dreadful situation. Certainly, the loss of humor and the sense of lonely suffering must have been present in the two police officers who committed suicide during the pressured early days of Katrina flooding.

Sublimation or substituting safe activities for desired activities was one coping method used by many survivors. When children could not go on with their normal activities, parents helped them create games in the Superdome that could be played with a minimum of equipment.

Prayers and songs were used in some family groups when little else could be done on rooftops as survivors waited for rescuers.

Postponing attention and suppressing wishes can be used when a situation is dreadful. Survivors had to ignore and endure the sights, sounds, and smells of bleeding, vomiting, diarrhea, crying, and all the physical and emotional reactions of those around them. First aid had to be rendered despite stomach-turning conditions. People had to see and hear things they had never experienced before and which they were helpless to change. This ability to tolerate and ignore things that can usually be avoided was described among most of the survivors interviewed by reporters and rescuers.

Among the unhealthy or ineffective coping techniques was blaming problems on others and distorting or exaggerating problems caused by others. These were among the first reactions used by virtually every single leader interviewed in the earliest days. Sometimes blame was accepted by looters who were caught in the act. But even those people didn't want to take all the blame so they said, "Well, we're not the only ones—others were doing it, too."

Some people, even those in very high places, have blamed President Bush with distorting facts. A group of 60 leading American scientists stated in 2004 that although science has helped America become powerful and its citizens prosperous and healthy, the "distortion" of scientific knowledge by the Bush administration for political purposes must end. They were referring, among other things, to the denial of environmental data as well as other scientific information.

A very poor coping method akin to denial is regressing from problems and letting other people solve them. That sort of shying away from problems is often used when one doesn't know what to do. It's like an ostrich sticking its head in the sand and becoming oblivious to what is going on. Perhaps many of the individuals who *chose* to stay in their homes in the Gulf Coast despite storm flood warnings were sticking their heads in the sand. Some had their reasons, such as trying to protect their premises from looters, or disbelieving the warnings.

Another poor coping technique is to intellectualize or rationalize one's own actions. It's almost like lying about what you did while making yourself sound reasonable and smart, when you were just plain wrong or stupid. Many leaders at all levels of government were guilty of this during the stress of dealing with Katrina. Michael Brown gave what sounded like a reasonable story that there were so many deaths because people ignored warnings to leave New Orleans. That sounded plausible until it became clear that many did not have a method of transportation yet wanted to be evacuated.

Perhaps the worst coping method is to deny that problems exist or to block information and acceptance of what the real problem is. Denial of fact is responsible for many deaths on the Gulf Coast and the devastation. However, people who denied real facts didn't start the day before Hurricane Katrina. The officials who denied facts began years earlier as will be shown in the next chapters.

The real tragedy unfolding in the Hurricane Katrina aftermath is the result of human nature. What else? Our human nature is what gets us into trouble and the only cure is learning to do better, and especially learning to handle stress better. For years, engineers and scientists have warned that the levee system protecting New Orleans needed improvement. Why didn't predictions of the destruction of New Orleans by a hurricane or floods move authorities to action, even though it would cost millions or even billions? Because nobody wanted to bite the bullet, spend the money, and make some people mad. They wanted to please all the right people and be liked.

President Harry Truman had it right—"If you can't stand the heat, get out of the kitchen!" Harry took a lot of heat for his decisions and if he were here today, he might add, "If you care so much about being liked, elected, and praised, you're no leader!"

Leaders aren't the only ones who have problems with their human nature, their reactions. The good and bad behaviors of victims, survivors, and rescuers in the streets and the Superdome have much more to do with human nature than with Mother Nature. We must examine the reactions of humans under stress which have turned this disaster into a tragedy and travesty.

New Orleans and Louisiana has been beset by problems of human nature for many, many years. It has a long history of political corruption, law enforcement corruption, and personal corruption. Did this tradition of corruption create indifference to the people who were being served? Did this indifference, compounded by the indifference of the federal government, kill New Orleans and its residents? We will examine how human nature has added to Mother Nature by increasing death and destruction in New Orleans and the Gulf Coast.

2

Who Can Control "Ole Man?"

○ ○
"There is on the globe one single spot, the possessor of which is our natural and habitual enemy. It is New Orleans."

—Thomas Jefferson upon learning that Spain secretly gave
New Orleans to France

The truth is, we're not as much in control of our lives as we think we are. That is why the weather still occupies about one third of the time on the typical news telecast. Despite that, we find ourselves surprised when a natural calamity occurs, as if it were not only unexpected but downright wrong. That is denial of facts!

There has always been a tendency to seek blame when catastrophes befall us. If people could find someone or something to blame, they could perhaps eliminate that cause and try to control the climate, or so they hoped. Prehistoric people tried to control the climate by offering sacrifices, prayers, rites, and rain dances to the gods. In the past, major climatic disasters were said to occur because of man's evil and his punishment by God or the gods. Such a story was described in the Bible about Noah and the Ark.

> And Jehovah saw that the wickedness of man was great in the earth, and that every imagination of the thoughts of his heart was only evil continually. And it repented Jehovah that he had made man on the earth, and it grieved him at his heart. And Jehovah said, I will destroy man whom I have created from the face of the ground; both man, and beast, and creeping things, and birds of the heavens…And the flood was forty days upon the earth; and the waters increased, and bare up the ark, and it was lifted up above the earth. And the waters prevailed and increased greatly upon the earth…And all flesh died that moved upon the earth, both birds, and cattle, and beasts, and every creeping thing that creepeth upon the earth, and every man…And the waters prevailed upon the earth a hundred and fifty days…Taken from *Genesis 6:5 to 7:24*.

In order to control the weather, Noah and his ancestors tried to appease God, just as prehistoric people tried to appease their gods.

The French quarter of New Orleans with its sex shops and year-round Mardi Gras atmosphere has often invited theologians to suggest that one day the "Big Easy" would be swallowed by the earth in some awesome display of the divine wrath. To be sure, many wonder if a good God would allow natural disasters to befall and kill innocent people. In truth, however, those who suffer the most in these natural catastrophes—the poor and the black—are usually those to whom life has already been unkind.

Some said that God sent Hurricane Katrina because a city-wide gay pride parade was scheduled for the next week. Some said that God had punished Biloxi because it was a gambling mecca and New Orleans because of its sexual immorality. Some say that God sees America as a land of abortions, homosexuality, and greed that should be punished by terror such as 9/11 or floods.

People have reacted to disasters in much the same way since the beginning of time. Pliny, the Younger, wrote about the eruption of Vesuvius on August 24, AD 79. He described the tendency of some to blame the gods for this eruption, accompanied by earthquakes, which destroyed and buried the towns of Pompeii and Herculaneum.

> We also saw the sea sucked away and apparently forced back by the earthquake: at any rate it receded from the shore so that quantities of sea creatures were left stranded on dry sand. On the landward side a fearful black cloud was rent by forked and quivering bursts of flame, and parted to reveal great tongues of fire, like flashes of lightning magnified in size...Soon afterward the cloud sank down to earth and covered the sea; it had already blotted out Capri and hidden the promontory of Misenum from sight...
>
> You could hear the shrieks of women, the wailing of infants, and the shouting of men; some were calling their parents, others their children or their wives, trying to recognize them by their voices. People bewailed their own fate or that of their relatives, and there were some who prayed for death in their terror of dying. Many besought the aid of the gods, but still more imagined there were no gods left, and that the universe was plunged into eternal darkness for evermore. (Carey)

Volcanoes and earthquakes can happen in or create mountains. Floods happen more on very low land. The Low Countries of Europe (Netherlands) were called "low" because they were below sea level and experienced periodic floods. This led to the legend of Hans Brinker, the little chap who plugged the hole in the dike with his finger. In 1228, a flood smashed the first primitive levees in the Friesland area, killing perhaps 100,000 people. Since the Middle Ages, the Dutch have struggled to remain above water. Around the 14th century, the Dutch built a system of canals and dikes capable of controlling internal waters. By the way, the locks of Spaarndam, near Amsterdam, are the oldest in Europe.

Sometimes the Dutch dikes were insufficient to control nature and climate. More than half the country is below sea level and despite strengthening, occasionally a dike breaks. Bruges was overcome by sand in the 15th century, and Antwerp gradually replaced flood-ravaged Bruges in importance. Venice, of course, was later to have problems with its foundations, and benefited by Amsterdam's example (Cardini). Currently, of course, Venice is sinking into the sea at a much faster rate than ever before, but expensive steps are being taken to save it.

Currently, Dutch flood control authorities fear they are gradually losing their centuries-old battle with the climate. Floods are growing more frequent and require more strategies from water management authorities. In 1995, rivers and their tributaries surged, prompting the largest evacuation of Dutch civilians since the 1940s. At the last minute, the weather improved and reinforced dikes were sufficient to protect villages from devastation. Sea levels will continue to rise as the earth continues to warm, and storms will drop heavier rains over the region. How will the Dutch meet this problem?

They plan to try a method of opening holes in the dikes during heavy flows, such as China did after the flooding of the Yangtze put millions of people at risk. In effect, the Dutch will flood plains and farms to protect cities, even though it may result in food shortages. They will also try deepening river channels to make more room for the river. They fear that just building higher dikes will create more danger if they break. Sacrifices have to be made and currently they believe it's safer to flood farms than to flood cities.

Can You "Read" the Mississippi River?

America's largest river is the main source of flooding in our lifetime. The Mississippi River was called "Ole Man" or "Old Man" in spirituals. Samuel Clemens (Mark Twain) could write accurately about the river in *Life on the Mississippi,* because he received his pilot's license in 1859 and piloted on it for several years. He described trying to "read" the Mississippi River and the dangers it posed:

> The face of the water, in time, became a wonderful book—a book that was a dead language to the uneducated passenger, but which told its mind to me without reserve…It is the faintest and simplest expression the water ever makes, and the most hideous to a pilot's eye…
>
> The sun means that we are going to have wind tomorrow; that floating log means that the river is rising; that slanting mark on the water refers to a bluff reef which is going to kill somebody's steamboat one of these nights; those tumbling "boils" show a dissolving bar and a changing channel there; the lines and circles in the slick water over yonder are a warning that that troublesome place is shoaling up dangerously; that silver streak in the shadow of the forest is the "break" from a new snag.…

New Orleans is built on a flood plain, but floods are not the only danger along the path of the Mississippi River. The geologically vulnerable Midwest area includes the New Madrid zone, one of the biggest earthquake zones in North American history, which is still active. The famous New Madrid earthquake began on December 11, 1811, and the earthquake and tremors continued for days. The New Madrid earthquake was felt over several states though few people were killed.

Fear of the tremors grew and in Tennessee, Kentucky, Mississippi, Arkansas, Illinois, Indiana, Ohio, and western Virginia where the quakes were most forceful, the Methodist church increased its membership from 30,741 in 1811 to 45,983 in 1812. This gain of over 50% of so-called "earthquake Christians" occurred at a time when the number of Methodists in the rest of the country increased at only about 1%. Obviously, many believed the earthquakes were God's wrath and they must get right with religion to survive.

The San Francisco earthquake of April 17, 1906, had many of the same sights seen in the Katrina and Rita hurricanes and floods of 2005. That earthquake and subsequent fire left 225,000 homeless. Two-thirds of the city was destroyed. An estimated 700 to 3,000 people were killed, and 28,000 buildings were destroyed.

From the author's (Holloway) own book, *American History in Song: Lyrics from 1900 to 1945,* some songs depicted reactions to the catastrophe. Survivors of the San Francisco earthquake sang the 1905 hit song *Wait Till the Sun Shines, Nellie*, substituting "Frisco" for Nellie. About the only good thing to come out of the earthquake was the realization that the automobile was something useful, instead of just a toy for fun. All the available autos were called in to carry the injured and the aged. With approximately one car per 800 people before the earthquake, car production skyrocketed once its utility was recognized.

Writer Jack London described some of what he saw in that terrible earthquake.

> San Francisco is gone! Nothing remains of it but memories and a fringe of dwelling houses on the outskirts. Its industrial section is wiped out. Its social and residential section is wiped out…On Wednesday morning at a quarter past five came the earthquake. A minute later the flames were leaping upward…There was no opposing the flames. There was no organization, no communication. All the cunning adjustments of a twentieth-century city had been smashed by the earthquake.
>
> The edict which prevented chaos was the following proclamation by Mayor E. E. Schmitz: "The Federal Troops, the members of the Regular Police Force, and all Special Police Officers have been authorized to KILL any and all persons found engaged in looting or in the commission of any other crime"…Surrender was complete. There was no water. (Carey)

London's description of the looting and the police attempts to regain order remind us of the looting after Hurricane Katrina. Fortunately, there was much less looting after Hurricane Rita, mainly because there was almost total evacuation of the devastated areas. However, the Mississippi has a reputation for being unpredictable and for changing the nature of those who live along it and have to deal with it.

"Ole Man" has long been known to be a blessing as well as a flood threat. It was an important border between the "civilized" and the "uncivilized" during Benjamin Franklin's time and an extremely important conduit of transport and travel, and so was its final port, New Orleans.

Who Wants New Orleans?

Thomas Jefferson considered the Mississippi River and New Orleans to be extremely important. He made the following statement when he learned that Spain, who owned New Orleans in 1801, was passing it to France through secret treaties. "There is on the globe one single spot, the possessor of which is our natural and habitual enemy. It is New Orleans, through which the produce of three-eighths of our territory must pass to market."

Jefferson wanted Napoleon to cede New Orleans to the United States so that we might avoid going to war with France over who controlled that port. Unaware that Napoleon wanted to sell all of Louisiana, Jefferson began negotiations with France to purchase New Orleans, which led instead to the Louisiana Purchase.

The Lewis and Clark Expedition left St. Louis from the Mississippi River in the spring of 1804, ascending the Missouri River, using various rivers and land to cross the west and to arrive at the Pacific in November, 1805, where they built Fort Clatsop. That part of the Louisiana Purchase north and west of the Territory of Orleans was, from 1804 to 1805, known as the District of Louisiana; from 1805 to 1812, as the Territory of Louisiana; and in 1812 the name was changed to

Missouri Territory. Meriwether Lewis came back from the expedition to manage the Louisiana territory before he killed himself. He once said that except for New Orleans, Louisiana was a "barren waste."

Engineer Robert E. Lee and the Mississippi

In 1837, Lieutenant Robert E. Lee, fresh out of West Point where he had learned engineering, was ordered by the U.S. Department of Engineers to make a survey of the Mississippi River. He was to recommend methods to save St. Louis from a flood as the main current of the river was deflecting to the Illinois side and cutting a new channel through the bottom lands.

So difficult was transport on the Mississippi around St. Louis that in low water season, all steamboats had to place their cargo and passengers on keelboats to be towed by horses along the shores to the head of the rapids. The steamer that Lee and his crew acquired quickly experienced navigation difficulties and crashed against rocks. He operated from the deck of the crashed boat as he and his men surveyed the river and its problems. It took him three months to complete his work and prepare his report.

He recommended that the two rapids be improved by widening the channels with blasting and suggested moving rocks which obstructed navigation. He also recommended placing dikes to deflect the strong currents and to force the main current back into its original channel by driving piles and constructing cribs and wing-dams. The eddies thus created caused a deposit of sediment to be made between the dams, which gradually filled up the place with solid material where the wash-out had occurred, and diverted the unsteady stream back into its original course.

It was the beginning of the permanent improvement of the upper Mississippi River. At that time, only a few log houses, traders' stores, and military posts existed on the shores of the rapids and for hundreds of miles above. Many years later, the commerce of the river required larger boats and progress could be safely made through the channel.

That commerce supplied the wants of the millions who have since made of the upper Mississippi and of the plains of the Red River of the North the granary of North America. The fertile land along the Mississippi had been used by prehistoric Indians for some 9,500 years, as they and subsequent farmers perfected the domestication of grasses, and harvested corn, wheat, rye, barley, rice, and sugar cane.

Many other cities sprang up after St. Louis, such as St. Paul and Minneapolis, and this was in large part thanks to the services of Robert E. Lee, as may be seen on a riverbank plaque in St. Louis. Of course, his leadership of the Confederate forces during the Civil War is well known.

The Mississippi attracted a lot of attention later when it began rising in the fall of 1926. By New Year's Day, 1927, it reached a peak and would remain in flood for over 150 days. The flood destroyed levees, inundating over 25,000 square miles of land, but somehow missed New Orleans. The U.S. Federal government had to intervene with funds, relief, recovery, and future plans.

In 1927, the flood of the lower Mississippi River displaced at least 700,000 people despite much river engineering. It was like Hurricane Katrina in that over 1,000 died. Some 2,200 buildings were washed away and refugees were living in tents by the levees. Herbert Hoover, Secretary of Commerce, organized relief programs. President Calvin Coolidge's quick dispatch of Hoover helped because he was given sweeping authority to coordinate all local, state, federal, and private endeavors. His success helped him to become president in 1928.

During the 1927 flood, blacks were held captive on the levees thanks to policy bungling intended to keep the labor force intact. The forced laborers were controlled by armed National Guard members. Meanwhile, movers and shakers in New Orleans tried to protect their investments.

Ultimately, the flood set up a precedent of vastly expanded federal involvement in local affairs described in John Barry's book, *Rising Tide: The Great Mississippi Flood of 1927 and How It Changed America.*

Barry also described the Ku Klux Klan, other organizations, the media industry, and Mississippi blacks in society and government. Barry recounts how political ineptitude and personal greed helped cause the flood's devastation, and how human behavior proved more destructive than the flooded river itself. That is the story of mankind. We are our own worst enemy. Or as one comic strip so sagely said, "I have seen the enemy and they is us!"

The Mississippi River in Song

Blacks at the lower economic level often composed songs about their predicament. Another song from *American History in Song* is relevant to how the flood affected blacks. The Red Cross was there to help out but Walter Roland recorded the demeaning treatment of poor blacks at some stores in *Red Cross Blues*.

> Said me and my good girl talked last night,
> Me and her talked for hours.
> She wanted me to go to the Red Cross Store
> And get a sack of that Red Cross flour.
> Say you know them Red Cross folks there,
> They sure do treat you mean.
> Don't want to give you nothin'
> But two-three cans of beans.

Blacks were at the bottom of the food chain, and were forced into near slavery by whites to build levees, but were the last to get help during the flood. Two months later, as the water began to recede, blacks began to leave for the north. Many songs about the flood arose such as *Mississippi Flood, The Mississippi Mud, High Water Blues, Since I Laid My Burden Down,* and *Rising High Water Blues.* The latter, by Blind Lemon Jefferson, described some scenes during the worst of the flood.

> People, since it's raining,

It has been for nights and days
Thousands people stands on the hill,
Looking down where they used to stay.
Children stand there screaming,
Mama, we ain't got no home
Papa said to the children,
'Backwater left us all alone'.

Charlie Patton's *High Water Blues* told how the blacks were kept from leaving the area by the whites as long as they were needed for work.

Lord, the whole 'round country
Lord, the river has overflowed.
You know I can't be stayin' here,
I've gotta go where it's high, boy!
I was goin' to the hill country
'Fore they got me barred.

The South had long taken advantage of blacks since the days of slavery, and their treatment during the flood of 1927 reminded them of their awful plight. Earlier, blacks had been going north at a pace estimated by the Labor Department in 1923 at 500,000 blacks a year. The song *Bye Bye Blackbird* by Charles Strouse and Lee Adams explained the migration thusly:

Pack up all my care and woe,
Here I go, singin' low
Bye bye blackbird
Where somebody waits for me,
Sugar's sweet, so is he
Bye bye blackbird.
No one here can love or

Understand me
Oh what hard luck stories
They all hand me.

Musicians often followed the news to attract bigger audiences. The biggest news story of the year, besides Charles Lindbergh's flight, was the Mississippi flood and its aftermath. The last week of 1927, *Showboat* (based on the Edna Ferber novel) debuted on Broadway. It was a new kind of musical by Jerome Kern and Oscar Hammerstein, II. It combined a serious racial theme with enjoyable entertainment, had a story line told in narrative and song, and had a cast of black and white personnel. And it dealt with the topical themes of the Mississippi River and mistreatment of blacks by whites.

The story was based on the persecution of a woman with Negro blood married to a white man. In one scene which shocked audiences then, the man cut the woman and sucked her blood to show that he accepted her, was like her, and had the same blood in his body.

The lyrics reiterated the treatment of blacks by whites along the Mississippi in *Ol' Man River.*

Niggers all work on de Mississippi,
Niggers all work while de white boss play,
Pullin' dose boats from de dawn to sunset,
Gittin' no rest till de judgment day.
Don't look up an' don't look down.
You don't dast make de white boss frown
An' bow your head an' pull dat rope
Until you' dead.

The production was praised *and* criticized by blacks and whites. Over the years, the words "niggers all work" were considered too objectionable and were changed to "darkies all work," then to "colored folks work," and even to "here we all work."

Another popular song from *Showboat* was *Can't Help Lovin' Dat Man*. It used the usual Negro stereotype of "lazy" to characterize the man she loved in these lines:

Fish got to swim, birds got to fly,
I got to love one man till I die
Can't help lovin' dat man of mine.
Tell me he's lazy, tell me he's slow
Tell me I'm crazy, maybe I know.
Can't help lovin' dat man of mine.

Showboat was not the only famous literary work to come out of Mississippi floods. Some 12 years after the 1927 flood, William Faulkner wrote a story about it called *Old Man* (interspersed with another story called *Wild Palms* about climatic conditions.) Just as in the 2005 Katrina flood, prisoners were released temporarily to help in the rescue effort, so did Faulkner's story depict a convict released to help find flood victims. The convict and protagonist in Faulkner's story rescued a pregnant woman in his boat, and they floated down the Mississippi, just as Samuel Clemens' Huckleberry Finn and Jim did.

The woman gives birth to her child, helped by this convict who has lived in prison so long that he knows no other life than as an inmate. His only wish is to get to town and give his charge over to others and return to prison, even though he could have escaped. However, he had principles and wanted to do it the right way—to find somebody he could give her to, and to find something solid to set her down on. He then proceeded to go back to jail.

Just as the Mississippi River has changed the character of some people, it has demonstrated the enduring characteristics of others. Wrestling this force of Mother Nature has shown our human nature at its worst and at its best. The goodness of man in helping others survive was demonstrated over and over again in Hurricanes Katrina and Rita in 2005.

3

Facts That Were Denied

○ ○

"There is a tide in the affairs of men,

Which, taken at the flood, leads on to fortune;

Omitted, all the voyage of their life

Is bound in shallows and in miseries."

—*William Shakespeare, Julius Caesar IV, iii, 217.*

One of the main problems creating the deaths and devastation of Hurricanes Katrina and Rita was the denial of facts by those who could have appropriated flood control projects. What facts were denied?

An excerpt of an important book by Pulitzer Prize Winner John McPhee was published in *The New Yorker* on 2/23/1987. *The Control of Nature* startled readers as it described the battle of man against nature in three areas: the U.S. Army Corps of Engineers vs. the Mississippi River in Louisiana; the Icelander vs. flowing lava to save a crucial harbor; and Los Angeles vs. debris flows from the San Gabriel Mountains.

In a *New Yorker* excerpt, McPhee described the monumental battle of engineers to prevent the Atchafalaya River from creating a new route to the Gulf that would cut off New Orleans and Baton Rouge from the rest of the United States. This painstaking researcher's predictions were revived after Katrina on 9/12/2005 when *The New Yorker* reprinted the excerpt. This phrase will give a glimpse of that incredible battle between human nature and Mother Nature:

> In each decade since about 1860, the Atchafalaya River had drawn off more water from the Mississippi than it had in the decade before…As the Atchafalaya widened and deepened, eroding headward, offering the Mississippi an increasingly attractive alternative, it was preparing for nothing less than an absolute capture: before long, it would take all of the Mississippi, and itself become the master stream…
>
> The consequences of the Atchafalaya's conquest of the Mississippi would include but not be limited to the demise of Baton Rouge and the virtual destruction of New Orleans. With its fresh water gone, its harbor a silt bar, its economy disconnected from inland commerce, New Orleans would turn into New Gomorrah.

The struggle to contain the Atchafalaya was only one part of the "control of nature" being attempted by engineers around New Orleans. In 1990, a federal task force began restoring lost wetlands surrounding the Crescent City. Wetlands had eroded exposing the city to storm surges which could be reduced if land was extended farther out to the Gulf. The preservation and enhancement of wetlands was too expensive for local or state funds to handle alone, so the federal government was involved in proposals. The Bush administration had earlier promised no loss of wetlands, but in 2003, withdrew financial support unless the project was "related to interstate commerce."

A Louisiana flood killed six people in 1995 causing Congress to create the Southeast Louisiana Urban Flood Control Project. In that project, the U.S. Army Corps of Engineers strengthened and renovated levees and pumping stations. Despite that improvement, city leaders continued to be warned by FEMA, the Corps of Engineers, and local engineers that the levee system was weak. However, comprehensive repairs were not approved because of lack of funds and disputes over what replacements were needed and where they were needed.

A *Time* magazine article by Joe Suhayda on July 10, 2000, described what could happen if a flood were to lay waste to New Orleans. He wrote that a Category 5 hurricane would cause Lake Pontchartrain to overflow, and pour down millions of gallons of water on the city. Then evacuation routes would become blocked, buildings would collapse, and chemicals and hazardous waste would turn the floodwaters into a lethal soup. He concluded that what might be left of the city would probably not be worth saving. Sound familiar?

But why is it worth saving? Why should people have paid attention to facts and predictions that were presented? What is the importance of New Orleans to the country? New Orleans is the largest port in the United States, the fifth largest in the world, and the port is paralyzed without a workforce. That workforce was largely located in the area that flooded out. Is it possible that the city, the state, or the federal government underestimated the importance of New Orleans and overlooked warnings to save money? Where and how can the workforce be returned to maintain the port traffic and activities so vital to U.S. commerce?

Warnings Were Clear and Strong

In 2001, FEMA warned that a hurricane striking New Orleans was one of the three most likely disasters in the U.S. FEMA was not the only source of a foreboding about hurricane damage in New Orleans.

Mark Fischetti wrote an article for the *Scientific American* called "Drowning New Orleans" which appeared October 1, 2001, and predicted the disaster of Hurricanes Katrina and Rita. Fischetti suggested restoring marshes by adding gates to certain river dikes to let floodwaters deposit sediment in the marshes, and enhancing the levee system to withstand a category 5 storm surge.

Fischetti's article was based on computer models that Louisiana State University had been running for several years. A plan had been put together in 1998 by scientists and engineers to protect New Orleans from a Category 4 or 5 hurricane that might come from the south as Katrina and Rita did. The warnings were there in Fischetti's article.

Some have said that New Orleans residents lost out to other White House priorities such as the tax cut to win friends and influence people and the Iraq war.

New Orleans Mayor Morial in 1998 during Hurricane Georges had to deal with the first evacuation of the city. A poll of the citizens of the city found that only 50 percent evacuated. About 20 to 25 percent found themselves in shelters such as the Superdome, the Convention Center, and another 25 percent refused to go. That poll should have let city leaders know that many would not leave.

What Do People Think of New Orleans?

It has been estimated that Louisiana produces one-third of the country's seafood, one-fifth of the oil, and one quarter of the natural gas. Perhaps most Americans felt the Hurricane Katrina crisis first at the gas pumps.

Is New Orleans, as some say, the future of Miami, New York, San Diego, and all coastal cities in the world where sea level rises could cause problems? Many say that we must start thinking about a new energy future.

In 2002, the Environmental Protection Agency submitted a study on global warming to the United Nations. It reported that climatic changes have global consequences for human health and the environment. Although the White House claimed that the report required no action on global warming, scientists presented much data on how the rising temperature of the oceans has produced more frequent and more severe hurricanes. The decision to do nothing about these actions has brought the Bush administration much criticism during the Katrina aftermath.

Hurricane Ivan exposed significant flaws in New Orleans' disaster plans in 2004. Many thought that even though New Orleans dodged a direct hit, Hurricane Ivan would cause the city's disaster plans to be improved, but it did not happen.

In 2004, the U.S. Army Corps of Engineers proposed to study how New Orleans could be protected from a catastrophic hurricane, but again funding for the study was unavailable as were funds to study how to hold back the waters of Lake Pontchartrain. Additional budget cuts at the beginning of 2005 forced the New Orleans district of the Corps to impose a hiring freeze. Although the Senate debated adding funds to fix New Orleans' levees, funding was still not approved.

The New Orleans *Times-Picayune* published a series of newspaper articles on the federal funding problem before Hurricanes Katrina and Rita. Their series included accusations that the Bush administration's policy allowed New Orleans to be vulnerable to a category 4 or 5 hurricane.

This series came out not long before the National Oceanic and Atmosphere Administration (NOAA) predicted a 95% to 100% chance of an above-normal 2005 Atlantic hurricane season, according to a consensus of scientists at the Climate Prediction Center (CPC),

Hurricane Research Division (HRD), and National Hurricane Center (NHC). This forecast reflected the highest confidence of an above-normal hurricane season since NOAA's outlooks began in August 1998.

Numerous books, reports, articles, and series by those who know the most about hurricanes and floods were virtually ignored—denied. It's as if a patient were told by his doctor that he had terminal prostate cancer and would be dead within three years unless he took a medicine which killed the male hormone—in other words, he could live a normal life except for no sex. How many would be foolish enough to do nothing just to remain macho? And there probably is some connection between those who ignore facts and assume that they, their city, and their administration are tough enough to handle whatever comes.

Mother Nature vs. Human Nature

The whole issue of how to control Mother Nature and whether we must heed warnings to move or secure our cities has gone on since the beginning of civilization and continues. However, each state in our country has its own particular problems.

Wildfires are usually the result of human activity. People create the potential for major disasters through the decisions they make regarding where to live, where to concentrate development, as well as the actions they choose or don't choose to reduce potential losses. Decisions to permit development in forested areas bordering cities mean that we can expect losses from wildfires in the future.

Wildfire management has become a major problem for the country. That vividly illustrates why the U.S. must continue to support an emergency management system capable of dealing with disasters of all types, whether caused by natural events, failures of technology, or intentional acts of terrorism. Those who build home sites near wildfire areas may be no different than those who build near other geologically fragile areas and low lands along the Mississippi River and the Gulf Coast.

Another geologically fragile area has only recently been discovered. A recent survey of a bulge that covers about 100 square miles near the South Sister volcanic mountain in Central Oregon indicates the area is growing. Geologists believe the bulge near Bend could be a volcano in the making or a major shift of molten rock under the Cascade Mountain Range. Recent eruptions have occurred at Mount St. Helens nearby in Washington. Oregon has some of the most active volcanoes in the United States, including Mount Hood, Crater Lake, Newberry, and South Sister.

There are many active volcanoes under the ocean near the Oregon coast. Recently, sensors have been used to measure the amount of seismic activity. Is Oregon to stop building near the ocean or over the Bend area where the bulge is located, or is the state to order people to evacuate the area because of possible catastrophes? These are hard questions but they must be considered by city councils, city planners, state authorities, climatologists, and geologists.

Sometimes natural disasters are not the only part of our war with Mother Nature. James Howard Kunstler wrote *The Long Emergency* about our coming oil and gas shortages. He warned that without reliable supplies of oil and gas, the luxuries of central heating, air conditioning, cars, airplanes, electric lights, inexpensive clothing, recorded music, movies, hip-replacement surgery, national defense, and modern life will suffer.

His book suggests that we don't even have to run out of oil to start having severe problems and that depletions could begin to occur in 2010. Even though he predicts that it will change everything about how we live, are we going to do anything about that now? Are we going to change our dependence on oil and gas significantly within the next five years? Are we going to heed his warnings, ask geologists and scientists to explore this scenario and listen to their findings and recommendations?

Hurricane Georges arrived in September 1998. Its ferocity scared the scientists, engineers, and politicians into reaching a consensus. Late in 1998, the Louisiana governor's office, the Department of Natural Resources, the U.S. Army Corps of Engineers, the Environmental Protection Agency, the Fish and Wildlife Service and coastal parishes published *Coast 2050*. This plan for restoring coastal Louisiana was predicted to cost $14 billion.

Louisiana members of Congress wanted to rebuild the coast primarily by diverting water and silt from the Mississippi River across marshes and rebuilding barrier islands. Promoters of *Coast 2050* said it would begin to reverse some of the losses of the past 100 years and restore natural hurricane protection. However, at best it was predicted to provide only partial protection from hurricanes. Even if the entire coast could be restored to where it was a century ago, large storms could still devastate the area with floods, rain, wind and tornadoes inland, said scientists and engineers. They believed, though, that additional fixes were needed.

After Katrina was so destructive, German environmental minister Jurgen Trittin accused President Bush of shutting his eyes to the "economic and human damage that the failure to protect the climate inflicts on his country and the world economy through natural catastrophes like Katrina," according to the *Frankfurter Rundschau*. While Trittin was accused of insensitivity, America's consumption of fossil fuels and catastrophes like Katrina send a message to other countries that America is indifferent to the dwindling supply of earth's resources and indifferent to human needs.

Can Insurance Premiums Keep Up With Losses?

The Association of British Insurers issued a 2005 report, predicting that because of climate changes, losses from hurricanes in the United States, Japanese typhoons and European storms may increase by 60% in the coming decades. National Association of Insurance Commissioners has found that premiums have not been able to keep up with rising losses due to climate changes in the United States.

Bush administration officials have countered that global warming has not been proven scientifically. Therefore, they say that the U.S. doesn't need to cut back on producing greenhouse-gas emissions or participate in the Kyoto treaty, an international effort to decrease these emissions. If they did participate, they would have to use expensive methods to limit those emissions. Some 140 nations have ratified that treaty.

Moreover, although human oversight and errors obviously contributed to Katrina's death toll and destruction, the prime mover was Mother Nature. While the forces of Nature played the most direct role in the inundation of New Orleans, global climatic changes are in part triggered by global warming. Global warming is partly due to the uncontrolled emissions of carbon dioxide and other "greenhouse gases" by human action, which comes more from American activities than any other countries.

The United States government bears some responsibility for denying the mounting scientific evidence of global warming and for impeding efforts by the majority of nations to slow the impact of global warming. The White House has joined a few other countries (Australia, South Korea, India, and China) to draft a different Kyoto Protocol. This would permit greenhouse gas emissions to continue. The U.S. claimed that it would "damage" the U.S. economy to go along with the 140 nations and the Kyoto Treaty. Of course, it would damage the profits made by those businesses that create gas emissions. The U.S. is the world's largest consumer of energy per person so its businesses would be damaged most of all.

Looking Death in the Face

The loss of lives during hurricanes Katrina and Rita may never be entirely accurate and now hovers around 1,300. The massive flooding triggered by Hurricanes Katrina and Rita may cause New Orleans to become the first city in history to be lost to global warming. Despite its marvelous history and culture, New Orleans was a disaster waiting to happen and many knew it and tried to let others know.

There is growing evidence of ice melt in the Antarctic icecap that warns of accelerated global warming. Hurricanes Katrina and Rita may cause the Bush administration to reconsider its energy and environmental policies and replace ostrich-like escapism with leadership to deal with the crisis of global climatic change.

By avoiding the challenges of global warming instead of coping with these issues, and by obsession with low price band-aid fixes instead of upgrading storm protection, our leaders will only worsen the damage and cost when Mother Nature comes calling again.

Other facts that were denied involve the knowledge of human nature and the studies of crowd behavior. Some leaders (mayors, governors) assume that people will panic if told that they are in serious danger. Why, despite research evidence, does the idea of "panic" capture popular imagination and continue to be evoked by leaders? Is there any scientific justification for the continued use of the concept? Panic has always been considered part of sociology and psychology.

What constitutes panic is illustrated by stories of disaster behavior in newspapers and journals. The oldest view equates panic with extreme and groundless fear. Another view describes panic as flight behavior. There is the assumption that flight will occur only if there is a perception of escaping a threat. Disaster researchers in particular have emphasized that hope of escape rather than hopelessness is what is involved. Persons who perceived themselves as totally trapped such as in sunken submarines or collapsed coal mines do not panic because they see no way of getting away from the threat. Discussions of panic assume that panic behavior is "irrational." Most people who believe there is a way out of a deadly situation would be irrational if they didn't try to escape.

Some say that panic behavior is very contagious, and that human beings are easily swept up into the behavior. It is as if participants are easily overwhelmed by fearful emotions and will trample others in their path. This happens on some occasions such as soccer games where bleachers collapse or some such thing, but that is relatively rare.

A study of reactions by Americans to a nationally broadcast radio show (Orson Welles) supposedly sounding like a documentary of an alien invasion from Mars showed that those few who panicked upon hearing the broadcast lacked "critical ability."

To this day, that study is cited as scientific support for this view of panic, even though the research has come under sharp critical scrutiny. Only a small fraction (12%) of the radio audience ever gave even any remote credence to the idea that the broadcast was an actual news story. And the accounts of flight behavior as well as other illogical and bizarre actions reported were taken from journalistic accounts of the time which reported that supposedly numerous Americans fled wildly to get away from the alien invaders. Close examination of the actual news report that appeared found that most of them were sensationalized anecdotal stories that these days are typically reported in the so-called tabloid press.

Studies of persons caught in potential panic situations such as a fire in a night club and a stampede during a rock music concert found that involved persons did not engage in animal-like behavior, contrary to what many early writers on panic suggest occurs. Instead of ruthless competition, the social order did not break down. There was much evidence for rational responses in the face of the crisis. While strong emotions were experienced, these did not lead to maladaptive behavior.

Perhaps the idea of panic is necessary to highlight the fact that people react remarkably well in most stressful situations and that the social bonds between and among people usually hold up. The media report the *absence* of panic as if the normal expectation is that panic will occur.

On the contrary, there is a sort of denial in the minds of many people—a denial that there is danger, and a denial that they must do anything about it. When a mayor tells a community to get out of town because a hurricane is coming, it may be very hard to get their attention. The rarer the event that is predicted, the more denial will be used. People want to disbelieve warnings, especially if warnings were issued earlier and storms were less serious than predicted. Once someone has warned that a wolf is coming enough times without the appearance of the wolf, the warning is ignored.

However, warnings to evacuate when Hurricane Rita (category 5 at first) was about to hit the Gulf Coast three weeks after Hurricane Katrina got the attention of the public and people accepted warnings and evacuated more quickly.

Men and Women Differ in Their Reactions

There is also some difference between men and women, of course. Vive la difference! It may save you if you work or live or talk to women, who are quicker to evacuate than men. Men are reluctant to appear cowardly. Running to avoid peril can seem unmanly to some. Just as men don't like to ask for directions because they hate to admit they need help, they also hate to admit that something is about to happen that they can't control.

Women have had to ask for help all their lives because they aren't as physically strong as men. Thus, women don't worry about showing cowardice if they flee a potential storm. It just seems like the smart thing to do. However, sometimes a woman lays her fears aside because her man promises to "protect" her, and she surrenders to his wishes in order to keep peace with him and let him look manly.

Reactions to Warnings

Doing nothing about a potential storm, of course, takes no extra time and thought. There are plenty of other things to occupy our minds already—our job, our children, our health, our debts, our friends, our activities. It has been hard enough for city planners to get people and builders to avoid dangerous areas or lowlands as they develop new home sites and those planning decisions occur when there is no rush and plenty of time to think over possible consequences.

People were surprised that so many stayed in New Orleans despite threats of floods. Yet, many of us remember an old guy named Harry Truman (like the President) who stayed at home when Mount St. Helens was predicted to erupt, and paid with his life. Those who have stayed at home when storms were predicted earlier and lived to tell about it, assume the same thing will happen again. They feel vindicated in going against orders. Who wants to uproot themselves to unknown and unpleasant circumstances, especially if money is an issue?

Those who study emergencies have advice for city leaders when they must convince a community of real peril. Leaders must make a warning very specific. They must be very clear about what to do—what action to take. They must state exactly who is at risk and where the target is, so that certain specific people will do something. The leaders must not be alone but must have coordinated the release of this information with other agencies so that everyone is saying the same thing at the same time—the mayor, the governor, the police, the fire service, the city council chairman, and the media.

Here is where the media is the most important. People must hear and see the same message from every direction—radio, television, newspaper, etc. When possible, this message follows on the heels of earlier information and education about the potential threat or disaster. There is little doubt that the media played the most important role in nearly complete evacuation of 3 million people before Hurricane Rita hit land three weeks after Hurricane Katrina.

The regular channels of communication and life support services like medical facilities are jeopardized in the event of a disaster. After Katrina hit, there was no prompt response of rescuers so people began to provide first aid to each other and to loved ones. When there is an emergency, people behave very well. They try to help each other, especially if they are with people they know. Even strangers help strangers in true emergencies as studies have shown countless times.

When there is widespread destruction, there are usually problems with communications and transportation. That makes it difficult for emergency agencies to find out what has happened where and to get there. Before other help can arrive, the survivors are there and can see what needs to be done. They are therefore always the first to help their fellow survivors.

Katrina survivors had to figure out by trial and error where to apply pressure to stop bleeding. It wasn't long before snakes and other creatures posed problems. Water moccasins were more frequent than rattlesnakes and a few bites had to be treated. Cardiopulmonary resuscitation and mouth to mouth respiration were used, often incorrectly, to treat drowning victims. Bandages for wounds and splints for fractures were fashioned out of whatever was nearby.

Attempts to rescue victims required manual carriers of every conceivable type, the use of knots, searching methods, construction of floating aids to rescue non-swimmers, and even rope bridge construction in one case. Victims had to be transported, food had to be found and prepared, shelters were configured, latrines were constructed, and water purification methods were a tremendous challenge.

Training in all these procedures is available to all residents in the nation under the (Community Emergency Response Team) CERT program administered by fire departments, but few people had undergone this training before Katrina. Will they after? There is nothing like drills to perfect skills, and just as leaders (first responders, mayors, governors, and presidents) need to practice for disasters, a family disaster plan is a good idea for everyone, but especially for those who live in vulnerable places.

In other kinds of incidents, survivors are the first to pick up the injured, put them in cars, and drive them to a hospital. However, cars could not operate on the flooded roads in the Gulf Coast after Katrina hit. There was no nearby hospital for survivors to walk to or be transported to. Everyone knows that the best thing to do for a victim is to get them to medical help but when none is available, the efforts of survivors to help each other using their own knowledge, common sense, and human nature was all they had and that was really pretty good.

Katrina was similar to 9/11 in that hospital triage areas were ready for the injured but almost no one arrived, or not for a very long time. In 9/11, most victims were dead. In Katrina, victims could not get to the hospitals that were standing ready because they were so far away.

The Superdome had been named as one of the main shelters in New Orleans. Most who could leave had driven or flown out if they had enough money and transportation to escape the storm. Some of those who could flee chose to stay in places they knew—their homes—rather than go to an unknown and unequipped place to stay for an indeterminate time.

If they decided to go to a shelter, they were instructed on radio and television about what to bring. Walking to the shelter was hard enough on some, let alone dragging supplies. So many chose to stay in their homes, damaged as they may have been, because they were still home in a familiar place.

Criminals and Emergencies

While most people do not take advantage of others in emergencies, it has been said that "criminals do not reform during a disaster." What is even truer, however, is the saying that "disasters don't turn ordinary citizens into criminals." Most survivors are concerned with other things than criminal activity during a disaster. That is one large difference from riots. Police will attest to the fact that riots occur when a community is torn apart, and people feel the need to take sides. However, in disasters, people come together to help each other survive.

Looting was reported but it was not always done by criminals. There are good reasons why citizens and police might loot in such a disaster. People need supplies to survive. They may break into a store for life-saving supplies, food and water, as well as a garage or hardware store for equipment or tools to assist in rescuing someone. However, when television reporter Diane Sawyer asked President Bush if we should draw a distinction in New Orleans between those taking survival supplies and those taking VCRs, he replied that there should be "zero tolerance" of looting for any cause!

Leaders who said they delayed making evacuation mandatory because they feared people would panic, loot, and hurt each other were not working with needed information. Leaders should know or should surround themselves with people who know more about human nature and how people have reacted to emergencies since the beginning of time.

Leaders often avoid leading and heeding facts because taking precautions is expensive, unpopular, or distasteful. For example, there were many warning signs that were ignored and denied before the 9/11 attack. Many terrorist attacks took place between 1992 and 2001, and even more were prevented by arrests. However, little was done to take precautions for an attack against the United States. Here are some of the events that warned us about Osama bin Laden.

In 1992, al Qaeda agents set off a bomb in a hotel in Yemen where American troops had stopped on their way to Somalia. In 1993, the World Trade Center in New York City was bombed. Also in 1993, men trained by al Qaeda attacked U.S. forces and shot down a helicopter in Mogadishu, Somalia, portrayed in the movie *Blackhawk Down.* In 1995, American diplomats were murdered by al Qaeda members in Karachi, Pakistan, in retaliation for arresting the mastermind of the 1993 World Trade Center bombing. Also in 1995, Osama bin Laden funded a truck bomb attack in Riyadh, Saudi Arabia, which killed five American servicemen. In 1996, bin Laden financed another truck bomb attack on a U.S. military base in Dhahran, Saudi Arabia, which killed 19 American servicemen and injured many more. In 1996, bin Laden signed a declaration (Jihad) to drive U.S. forces out of the Arabian Peninsula.

In 1998, bin Laden issued an edict for Muslims to kill American civilians and soldiers wherever they could. In 1998, al Qaeda attacked U.S. embassies in Nairobi, Kenya, and Dar es Salaam, Tanzania, with truck bombs. At the beginning of 2000, an Arab trained by al Qaeda was arrested for plotting to bomb Los Angeles International Airport during the millennium celebrations. Later in 2000, bin Laden funded al Qaeda to bomb the USS Cole in the Yemeni port of Aden, killing 17 sailors and injuring 39 others.

In 2001, the Phoenix F.B.I. notified headquarters that there were many Arabs taking flight training. These warning signs did not put the U.S. on sufficient readiness for the attacks on 9/11/2001. Somehow, the Bill Clinton and George W. Bush administrations denied that the U.S. could be attacked.

However, when the attacks occurred, human nature led people to do things very well despite confusion by officials. Officials in the second tower were telling people not to panic. They were still assuming that an airplane accidentally hit the first tower, just as an airplane accidentally hit the Empire State Building in World War II. Fortunately, most people ignored the advice and fled out of the tower, thus surviving. That was similar to the passengers of the flight that crashed. They learned on cell phones about the other attacks on the Twin Towers and the Pentagon. They decided to prevent their plane from being used as a weapon on another important building and they worked together, knowing that they would die.

While individuals behaved well during Katrina and Rita events, organizations had their problems with communication, transportation, and loss of command posts and key personnel.

4

Police Had Two Roles

o o o o o o o o o o o o o o o o o o o o o o o o o o o o o o o
"Of all the tasks of government, the most basic is to protect its citizens against violence."

—Secretary of State John Foster Dulles

On the fifth day after Hurricane Katrina, New Orleans police officer Paul Accardo, 36, sat in an unmarked police car and shot himself to death. He was one of two New Orleans police officers who committed suicide within a week after Hurricane Katrina hit the city. He had been a spokesperson for the police department, reporting on incidents, but not working the streets or seeing sights as often as other officers saw them. He was an orderly guy who dressed neatly and kept a neat workspace. The other fellows reported that he had a little trouble when guys would pull a joke on him, and lacked a sense of humor that could balance a tough job and keep a good perspective on the ups and downs of life.

His supervisor, Captain Marlon Defillo, told reporters that Accardo committed suicide because he lost hope that order could be restored, and didn't know how to live with disorder. On those last days of his life, he saw people dead and dying in the streets, and saw the thousands awaiting perhaps impossible rescue from the Superdome. Defillo said that Accardo wanted to help women, children, and animals, but could do nothing to change the chaos. He took himself too seriously and thought he was supposed to rescue and do more than was humanly possible in his situation.

Accardo's home had been lost in the flood but so had many other officers' homes. He had gone without sleep but so had other officers. Some officers had abandoned their jobs but the majority continued to work despite terrible burdens. Other officers kept telling Accardo to cheer up because this was only temporary. Apparently, he couldn't believe them and despite having a wife, he ended his life. There is no greater message than to realize the importance of balance and perspective. Order and duties are important but we have to forgive ourselves if we can't do everything that is asked. There must be an appreciation of how insignificant we are, and that we can only do what we can do, and no more than that.

New Orleans Deputy Police Chief W.J. Riley identified another officer who committed suicide as Patrolman Lawrence Celestine, who also used his own gun.

Several dozen of the city's 1,600 police officers failed to report for duty, and some turned in their badges, but most officers were on duty endlessly and selflessly. Published reports put the number of those who went AWOL as high as 200, but Riley declined to comment on those figures, saying more than 100 officers may have been trapped in their own homes or unable to reach command centers.

On top of the burdens of law enforcement, officers had to forage for food and water to survive and even for places to relieve themselves. Police Superintendent Eddie Compass told reporters that his officers went into stores to feed themselves in some cases. They also had to deal with personal losses. And what affected most of them was that they didn't know where their wives or kids were. Many no longer had uniforms to wear and were in jeans and T-shirts. This produced some confusion for citizens who were unsure of whether those who were issuing orders had the authority to do so.

Compass resigned one month after Hurricane Katrina hit. He was being overwhelmed with criticisms, torn by the multitude of problems and directions from every possible source, and faced the investigation of some 250 officers for being absent from duty in the early days after Katrina.

The Two Basic Roles of Law Enforcement Officers

Usually, there are two basic roles that policeman play; either they are a protector or they are an enforcer. When they protect, they work among the people to listen, to hear what is needed, to see wrongs, to serve as the eyes and ears of police executives, to prevent crime from happening, and to help, rescue, and save citizens when possible. When they are enforcers, they must respond quickly to problems that have already occurred. Their job is punitive because they must arrest lawbreakers.

Within three days after Hurricane Katrina came ashore, New Orleans Mayor Ray Nagin ordered the police to stop their protective role of searching for victims and return to their enforcement role of stopping looters who had become increasingly violent. He told reporters that looters were getting closer to populated areas and had to be stopped right away. Some police said they were being shot at by armed citizens, and although their first priority was saving others, they had done little about looting until the mayor changed his orders.

In New Orleans, several different factors began to operate in the aftermath of Katrina. There was little ability of police to travel the area by car, to communicate adequately enough to call in off-duty personnel, or to set up command posts. Individual officers became unsure about whether to expect that their replacements would show up at the end of shift time, and had to work indefinitely without rest or, at times, access to food and clean water. They were worried about their own family members, homes, friends, and were sometimes torn between serving strangers or loved ones.

Sometimes they were perceived as enemies trying to move people from one place to another, instead of being seen as rescuers and helpers. Some were reported to allow looters to operate without apprehension. If their goal was to transport people to safety rather than to enforce the law, that was the more important mission as long as no danger was imposed by looters. Quick changes between the role of protector to enforcer is not always easy for police.

Some citizens armed themselves and fired at police, rescuers, and service people because they saw them only as enforcers who would arrest and disarm those in their way. Would things have been different if New Orleans officers had a history of more community-oriented policing in the poorer neighborhoods, being seen as protectors instead of punitive enforcers?

The New Orleans Police Department was called the nation's most corrupt police force during the 1990s. The city had the highest or nearly highest murder rate in America in several recent years. Many officers had been convicted of crimes including murder, rape, and robbery. The chances are that New Orleans citizens, and especially those who had frequent run-ins with the police, viewed law enforcement officers as the enemy rather than the helper. Modern policing suggests that police, firefighters, and emergency workers should constantly show their dual role, as both protectors and enforcers, in their daily work.

Problems Caused by Overeager Officers

We mentioned earlier that one lesson learned from the disaster of 9/11 attacks was that fire and police personnel deployed themselves, without waiting for invitations or orders, and arrived in New York at the Twin Towers creating problems for those who didn't expect them. Many agencies deployed themselves to New Orleans after Hurricane Katrina and wound up surprising authorities who didn't know how or where to assign them. Police and fire personnel who arrived from other cities were shocked to find citizens shooting at them. Furthermore, some who arrived were unable to help because they had not brought the right equipment for the situation they faced. For example, Phoenix firefighters took dogs trained to rescue people, but without solid ground, the dogs were unable to scamper about and find survivors. Other rescue workers arrived with vehicles that could not travel over flooded streets.

Emotions always operate at a peak in emergencies where clear lines of authority and communication are blurred. People get mad when others they don't know (often not even in uniforms such as FEMA) tell them what to do. People cannot work around the clock and make good decisions when they are exhausted. When relationships between workers have not been built up over a long time, it's too late to swap business cards at an incident.

New Orleans Fire Chief Charles Parent and his department operated from four staging areas and had many of the same problems as the police services. Fire departments began to arrive from New York, Illinois, Maryland, Texas, Arizona, and many other states. While their services were appreciated, they were often left with no assignment because the circumstances were so very unusual that they could offer no real help.

The International Association of Fire Chiefs met in August 2003 in Dallas and concluded that the training put in place after 9/11 prevented chaos during the 2003 electrical blackout on the East Coast. In a preventive and proactive role, police and fire personnel had prepared for emergencies by helping buildings appoint floor wardens and have power emergency kits available. They used their protective role to educate, listen, plan, and serve as role models. As a result, people were helped by wardens to find their way out of buildings, there was little looting, traffic soon began to move again, and people helped people. However, the New Orleans disaster did not permit officers to move about except by boat, and there was a severe shortage of boats.

Another lesson learned from 9/11 was to have interoperability of communication devices so that various responders (fire, police, FEMA, mayors) could talk to each other quickly. Police and fire personnel had also learned from 9/11 that they must be trained about what to do when they are deprived of communication and direction from the top. By the time of the blackout in 2003, they had a protocol about what to do if they couldn't talk to headquarters or to their supervisor. Part of the chaos during 9/11 was the lack of communication between services.

After Katrina hit, cell phones, land lines, and other communication devices were inoperable just when coordination between the various agencies was the most crucial. Local officials couldn't reach state officials, employees within agencies couldn't reach each other, and police officers couldn't reach each other, often simply because they didn't have radios. One communications executive commented that the New

Orleans Police Department never had a proper communications system, and their big tower used for radio communication in Jefferson Parish went down during the storm.

In some places, ham radio operators relayed messages back and forth between agencies. In other places where roads could be used, executives simply drove from one city to another to ask for or offer food, water, and supplies. Within a few days, the federal government brought in satellite phones but different frequencies caused the same problem noted in 9/11. Interoperability was needed so that all agencies could talk with each other on one system. Since then, federal money is pouring into the state to solve the communication (interoperability) problem.

Verbal and eye to eye communication is and was important because many service personnel must work together in emergencies. People must be treated with respect at incidents because everyone wants the same things. If service people are trained, ready, and have their equipment at 100% readiness, they can help. Young willing people who don't know what to do or don't have the same training can't help, no matter how eager they are.

When Things Break Down

What about salaries for city employees, as well as employees of other companies in this disaster? When there was no office for city employees to collect checks within two weeks after the disaster, city employees were being asked to register with a tracking service so that they could have direct deposit of salaries arranged by the city. Private corporations and other employers had similar plans or no plans. Two cruise ships docked at the Mississippi River with occupancy for about 2,000 city workers and officials.

While most of New Orleans was still under water, the U.S. Army Corps of Engineers tried to stop up broken levees with giant sandbags and concrete barriers. Meanwhile, the mayor called for a total evacuation of the city because of sanitation issues. He predicted a death toll of up to 25,000 because he ordered that many body bags. Other problems that plagued the city were the lack of mail service, the lack of public transportation, the lack of water available to the fire service to fight fires, and the lack of 9-1-1 services to be able to serve the entire afflicted area.

Authorities could not promise that New Orleans' public water system would be fully operational for at least three months after Katrina. At first, homes with running water had untreated Mississippi River water. Authorities warned that the water was unsafe and that dark water had up to ten times the amount of dangerous E. coli in it. FEMA set up plans to provide temporary housing for some 200,000 hurricane victims for up to five years in trailer homes. Insurance experts estimated losses at $40 billion and some put the total economic damage at more than $125 billion. Those figures rose after Hurricane Rita.

Back to Normal: Turf Battles

Gradually the law enforcement services began to be able to render service to the inhabitants again. Although protection of the people should be the common goal of all first responder groups, it can become difficult to carry out. There is the continuing problem of turf battles between politicians, press, people, pressure groups, police, fire, FEMA, and too many law enforcement agencies to promote coordination and harmony.

Edward J. Tully, Executive Director of the Major Cities Chiefs Association, wrote in an article in May 2002 about turf battles called "Terrorism: The Role of Local and State Police Agencies."

> The major flaw of some federal law enforcement agencies lies with their arrogance towards other federal, state, and local agencies. Unfortunately, there is little justification for the attitude. One of the most serious ramifications of this mind-set is the failure of the federal law enforcement agent to treat the capabilities of other agencies with respect.

In New Orleans, additional regional, state, and federal systems, social services, Homeland Security Department, as well as jails and prisons had to work together because even prison inmates had to be moved to safer places. One or two desperate criminals escaped and made headlines.

Most concerned police executives voluntarily cooperate in spite of the fragmentation between agencies. But there are islands of isolation where there should be bridges of understanding. The uniformed services is a group all bound to the same common public trust of protection for the citizenry but divided by a lack of trust in each other. This lack of trust is a natural defense that has developed in a time of change and financial cutbacks. Chiefs, sheriffs, other law enforcement executives covet information, rather than share information. Citizens see the police as one body of people, whether the uniform be brown or blue or white, cooperating to fight crime and prevent accidents. The public does not understand overlapping and fragmented policing service.

Law enforcement officials, federal officials, governor's office officials and others must work together to resolve friction over who has jurisdiction, problems which slow down response time in emergencies. Agencies must not allow the erosion of trust to continue—they must share information, make the best use of scarce manpower, and complement each other's efforts.

The primary goal during a major catastrophic event is public safety. The perception of danger can make a community fearful. Local police leadership must act to allay fear and emotions. The confidence a chief exudes in responding to and controlling situations will affect how safe citizens feel and how they react.

The International Association of Chiefs of Police recommended how law enforcement leaders should try to manage emergencies (and terrorist events.) They recommended:

- Plan before an incident occurs
- Develop policies and procedures

- Train personnel
- Rehearse possible events
- Acquire equipment and communications
- Establish mutual aid agreements and multi-jurisdictional teams
- Test public utilities such as electric, water, gas, waste treatment systems regularly and plan alternative sources as contingencies
- Set contingency plans and include back-up generator systems
- Help groceries, banks, and hospitals develop contingency plans to be prepared for emergencies
- Release accurate and immediate information to the public
- Identify where and how to access food, shelter, aid, transportation, and civilian assistance
- Train and rehearse communications between fire, police, EMS and other service providers
- Train and rehearse summoning help from governmental resources and politicians
- Stress must be expected and minimized
- Personnel must share objectives under the supervision of only one person—unity of command
- Law enforcement personnel are highly visible during crises and must use correct procedures
- A single spokesperson should be designated to deal with the media in order to correct rumors and give important information to the public
- The agency head, however, must speak if the community is to be reassured. The chief's presence and his manner are critical factors in community healing
- Reduce workloads for exhausted personnel

- Recognize heroic actions by those at the scene

- Act as role models in critical incidents

- Use key civic leaders to help control rumors. Strong relationships over a long time period will pay off when they provide help during crises

- Avoid unnecessary delays in sensitively delivering death notifications and the release of victim remains to families

- Mental health services should be made available for victims and responders

- Discuss emergencies and plans for emergencies with political leaders before incidents occur.

When we wrote the book, *Who Killed Detroit,* we described how Detroit has never recovered from those fateful days and nights of riots in 1967. Los Angeles has struggled with race riots and race problems as has Toronto, Newark, Minneapolis, Hartford, Dallas, and other cities that have also seen massive decreases in population, a loss of businesses, a decline of economic vitality, and slums developing in the downtown area. We concluded that other cities should beware because if we don't pay attention to history, it may repeat itself.

Former Detroit Police Commissioner Johannes Spreen came up with who and what was responsible. He thought the police, politicians, press, people, and pressure groups all had a hand in it. The police can certainly play a role in killing a city. Detroit's homicide rates make it America's deadliest big city. New Orleans is not far behind.

Louisiana and Mississippi have a long reputation for political corruption. Mississippi has the largest number of elected officials (per capita) convicted of crimes and Louisiana comes in third. The apple doesn't fall far from the tree, so with corruption at the top levels, it is little surprise that more than 50 Louisiana police officers (mainly in New Orleans) were eventually convicted of crimes in recent years and some are still serving time in prison or are on death row.

Cops often face the street code of violence, then silence. The code of silence occurs when people are reluctant to speak out, fearful for their own safety, and don't want to get involved. The other side of the code of silence is when police officers don't report on each other when they see wrong being done.

Spreen always thought that chiefs and supervisors should be models of integrity for their employees. This must start with attention to hiring. Psychological and background examinations must be connected with proven predictors of good performance and bad performance to select the best applicants. Once hired, carefully selected people must do Academy and field training, and ethics training should be included along with other subjects.

He believed that promotions, discipline, and terminations must be made fairly, consistently, and objectively, based on performance instead of "good old boy" favoritism. There must be an open policy requiring officers to report each other's serious misconduct and violations of policy standards. It must be clear that the organization will protect those who report misconduct.

Furthermore, he believes that the highest level of administration must visibly support these policies. We must be willing to have ourselves and our agencies open for inspection whenever there is any question about corruption.

Mayor Nagin's New Start with Police

The police profession must strengthen its relationship with the people of the community they serve. Police cannot solve crimes or protect the neighborhoods if people aren't talking to them. Yes, police must do more, do better, to win the hearts and minds of a community. And they must do more to root out their own bad apples lest the whole bushel become rotten.

Ray Nagin was elected as the Mayor of New Orleans on the platform of ridding the city and the police department of corruption. He understood that the person at the top of law enforcement agencies must be a good role model for his men. The chief's principles must show in the decisions he makes. In large departments, officers may have little contact with the chief and may learn more about what he stands for from in-house word of mouth and the press than from their contact with him. Good police officers and administrators abound, thankfully, and many examples of their good works can be given but none can wipe out bad actions by others.

Researchers have found that there are stages in the growth of organizational corruption.

1. First, officers perceive that there is administrative indifference toward integrity.

2. Next is the perception that obvious ethical problems are ignored as leaders intentionally look the other way or even cover up misconduct.

3. The third stage is the growth of fear as officers perceive that to survive as a leader, one must abide by the unwritten rules of internal politics. This stage may be accompanied by bitterness, officers rationalizing unethical acts in conversations with each other, and the hopeless conclusion that "everyone else is doing it." Kind of sounds like those looters, doesn't it?

4. The fourth phase is the survival of the fittest. The code of silence prevails when administrators hide misconduct rather than try to resolve it.

Deep within us, we assess the behavior of individuals, agencies, institutions, and societies by the measuring stick of integrity. Law enforcement must have integrity if it is to be acceptable to the people.

Today's chief must be all things to all people. A chief must be a leader, decision maker, confidante, politician, disciplinarian, therapist, mentor, administrator, taskmaster, spokesperson, community leader, educator, change agent, facilitator, partner, negotiator, role model, steward, student, parent figure, visionary, manager, minister and leadership developer. Of course, chiefs can't be all these things and some-times unions don't want him to be all these things. One could say the same thing about most high level leaders, including mayors, governors, and the president.

The chief must speak out on behalf of his employees and care what they want and how they feel. A chief must show good character through fairness, respect, dignity, and compassion. He must consider what is right rather than who is right.

The chief must demonstrate trust and fairness. A chief must avoid playing favorites, give credit to others, treat employees with respect, give employees the freedom they need to do their job, listen to differ-ent opinions, treat others as they want to be treated, value individual diversity, support and encourage people, and give fair performance feedback.

The chief should also develop good relations with community lead-ers, the mayor, the city manager, city council members, and others. Chiefs should also treat employees just like community-oriented police treat citizens, as customers who need respectful service.

Responsibilities of a Police Department

Police must become proactive. They must establish good relationships with residents. People know what is right and wrong in their commu-nities. Police must get them to share that information. Today most policing is based on simply responding to emergency calls, but that depends on residents reporting crime and other incidents to police.

The police must bear the blame in many cases for ruining relations with citizens if they use unnecessary force, intimidate and insult citi-zens in unprofessional ways, and don't take the time to offer help and protection. Part of this is because they work with some paranoia, fear-ing that they are seen as the enemy. Therefore, they sometimes act as aggressors before they are acted upon. That may be because they work daily with the worst people in our society, and they see the results of terrible crimes that can make them lose faith in the goodness of man.

Also, part of it is because some individual officers have become corrupt and seek personal gain or vengeance.

The key to changing the relationship between citizens and police lies more in the hands of officers than it does in the hands of citizens. No matter where it comes from, the message to improve relations in a positive direction must come from the top. The top means not only the top cops, who should be role models, but the entire police leadership.

One of the most important aspects to improving law enforcement services in a largely black community is to have blacks at the highest levels. The last two mayors of New Orleans had utilized black police chiefs. That was not true with the Sheriff's Department. For almost 30 years, the New Orleans Sheriff's Office and Prison was run by white Sheriff Charles Foti who had the reputation of being a corrupt racist, and was said to be far behind the times in running the prison.

In 2004, he left his position to become Louisiana's Attorney General and is currently pursuing charges against the owners of a nursing home where several patients died during the Hurricane Katrina. Sheriff Foti was followed by Marlin Gusman, the first black sheriff in the New Orleans area.

Criminals Take Advantage of Disasters

Finally, the FBI warned that we should be wary about criminals lurking on the Internet posing as organizations to help hurricane victims. The FBI has reported that more than 4,000 websites have arisen seeking donations for Hurricane Katrina relief efforts. It is unknown how many of these sites are legitimate and how many are fraudulent, but law enforcement and crime prevention experts are cautioning donors away from sites and organizations that are not widely known.

While donations for disaster relief are encouraged and welcomed, donors shouldn't give to organizations they've never heard of before. Online donations also typically require entering your credit card or bank account numbers, which is dangerous when dealing with an unknown entity.

People were warned not to reply with financial information to unsolicited e-mails. Fraudulent e-mailers have cloned the logos of many legitimate organizations and banks. In general, according to the FBI, if you didn't initiate the contact, don't continue it and don't reveal personal or financial information to anyone.

After Hurricane Rita, Sheriff Jack Strain in St. Tammany Parish promised to protect hurricane victims as contractors arrived to vie for reconstruction jobs. He announced that contractors doing business in the parish would need occupational licenses, work permits and a special permit from his office. He allocated five deputies to patrol damaged areas and check for proper permitting and licensure from contractors working in those communities. This was in addition to the scores of deputies on regular patrol who are concentrating efforts in those areas. He added that he would not tolerate criminals who take advantage of this calamity for personal gain.

It is a shame that law enforcement officials must stretch their limited manpower to dealing with various criminals while trying to help and protect victims of disasters.

5

Politicians Had Trouble Leading

o o o o o o o o o o o o o o o o o o o o o o o o o o o o o o

"Ambition is so powerful a passion in the human breast that however high we reach, we are never satisfied."

—*Niccolo Machiavelli*

President George W. Bush is not responsible for Hurricane Katrina. Who is to say that if his administration had approved more funding for flood control, it would have been applied in time, done correctly, or done where needed? Nevertheless, he has become a target for criticism, just as all leaders become targets when something goes awry during their term of office.

If people seek the highest office because of their own personal ambitions, they are sometimes perceived like a parent who is to minister to their children, the citizens. What child does not consider blaming a parent when he goes unprotected and suffers mightily? What child believes that his parents think only of their children and the welfare of those children? What child does not want to criticize a parent when he believes he is right and that parent, despite more knowledge, experience, and age is wrong?

Famous reporter and commentator Walter Lippmann bade farewell to his National Press Club colleagues in 1967. He was quoted in the *New York Times* on May 26, 1967, saying:

> I would have carved on the portals of the National Press Club, "Put not your trust in princes." Only the very rarest of princes can endure even a little criticism, and few of them can put up with even a pause in the adulation.

Thus, it is all the more surprising that on September 12, President George W. Bush accepted responsibility for his role in the federal response to Hurricane Katrina, with the statement quoted in Chapter One. Throughout his five years in office, Bush has not been one to admit mistakes. In the face of accusations that his administration made mistakes by invading Iraq on false assumptions, he has not been willing to accept much blame. Whether his acceptance of blame for mishandling Hurricane Katrina is sincere remains to be seen.

Bush seemed to learn things quickly between Hurricane Katrina and Hurricane Rita. Mindful of criticism that the federal government was slow to respond to Katrina, Bush jumped quickly on plans for evacuation of those in the path of Hurricane Rita. He had planned to visit Texas to review the Rita response three weeks after Katrina. However, he made a wise decision and cancelled at the last minute to avoid slowing down the preparations. That allowed authorities to focus on the safety plans for residents instead of the safety of a visiting president.

Three days after Katrina, New Orleans's Mayor Ray Nagin appeared to be accepting responsibility to stay and handle things rather than to flee for his own safety. He remained behind other evacuees at a makeshift command post in the Hyatt hotel, across the flooded street from City Hall. Peter Carlson of the *Washington Post* wrote about him on September 2, 2005, that he remained behind, like a captain determined to stay with his sinking ship.

Even though flood prevention was not a major issue in the election, he told a reporter in 2004 (*New Orleans magazine*) that his favorite book about the city was John Barry's *Rising Tide: The Great Mississippi Flood of 1927 and How It Changed America.*

Before his election, it had been refreshing to hear the Mayor speak about delivering the city from corruption. Louisiana and New Orleans have a long, well-known reputation for corruption. Billy Tauzin, former congressman, made the famous comment, "Half of Louisiana is under water, and the other half is under indictment."

Mississippi and Louisiana have been in the top three states for the number of elected officials that have been convicted of crimes. In 1991, Edwin Edwards ran for governor against former Ku Klux Klan leader David Duke, but despite his win, was sentenced to ten years for taking bribes from casino owners, and Duke was sentenced to prison for tax fraud.

How Do Mayors Decide Upon Evacuation?

Some have criticized Mayor Nagin because he and city officials waited so long to issue an evacuation order in a city where they knew that many had no car, no money for transportation, no way to afford hotels, and therefore no way to escape the storm and potential flood.

Nagin may have to do his own soul-searching in addition to the searching of critics. He may have been afraid to order mandatory evacuation because he feared the criticism of businesses, tourists, and citizens who would lose much money by closing up and fleeing.

Criticism is very hard to bear, and politicians, like most other people, do everything possible to avoid negative criticism. The fear of criticism is often what keeps people from taking the lead in doing the unpopular or issuing orders that will be disliked. In the end, criticism for a politician can mean that he loses his place and his prestige in the next election.

There are other reasons that government officials are slow to issue mandatory evacuation orders. There is a risk to those who try to evacuate, especially in the crowded conditions of an eminent hurricane. After Hurricane Katrina, officials were quicker to issue evacuation recommendations.

Two Louisiana residents fleeing Hurricane Rita were the first victims of the hurricane in Mississippi. The elderly couple was killed in a one-vehicle accident on U.S. Highway 98 in Pike County. They had fled from Lake Charles. A third passenger in the vehicle had only minor injuries.

The next victims of Hurricane Rita were from a busload of 37 nursing home patients headed from Bellaire in southern Texas to Dallas. The evacuation of the nursing home was not mandatory, but Bellaire officials had strongly urged all nursing homes in Bellaire to evacuate their residents. Firefighters helped put patients on the bus because it had no ramp for wheelchairs. Over half of the patients couldn't walk and many couldn't communicate well. The driver, a Hispanic, didn't speak English, so most of the patients couldn't communicate with him, but six health workers rode along with the patients. On the way, some passengers ran out of oxygen and they stopped to get more. Then the bus had a flat tire on the road.

Fifteen hours after it began, the bus had reached the Dallas suburb of Wilmer-Hutchins about 7 a.m. Motorists following the bus noticed smoke coming from the bus and signaled the driver, who pulled over. Others evacuating south Texas stopped their cars to help take patients from the bus. Emergency crews risked their own lives to rush in and help pull out hysterical elderly and disabled passengers.

Momentarily, the flames made contact with an oxygen tank causing an explosion and the bus quickly became an inferno. Twenty-three people died as flames devoured the bus. Their charred remains were found in the middle of the bus. Those in the front were able to escape through the door; those in the back were pulled out through windows smashed by rescuers.

Officials decided to load the burned bus onto a trailer and transport it away to make room on the highway for Hurricane Rita evacuees. Those taken to hospitals were treated briefly and released. The cause of the fire remained unknown.

Another early victim of Hurricane Rita was one person who was killed in Mississippi when a tornado spawned by the hurricane overturned a mobile home.

Later rescuers pushed their way into Beaumont neighborhoods once Rita receded along the Texas-Louisiana coast. The bodies of five people were discovered in a Beaumont apartment. A man, a woman and three children were apparently overcome by carbon monoxide from a generator they were using after the hurricane knocked out the electricity over the weekend. Another couple was found dead in their home from the collapse of a tree onto their dwelling. The death toll for Hurricane Rita (directly and indirectly) is currently in the 30s and will probably not reach 100.

An additional reason that the Mayor of Houston was reluctant to issue mandatory evacuation orders was because it meant moving some people who had evacuated New Orleans under terrible conditions in Hurricane Katrina only a few days earlier. It was very hard on them and reporters recorded their reactions.

"This reminds me of the Israelites marching in the desert," said one man as he waited for transportation to take him from Houston, Texas, to Fort Chaffee, Arkansas. He was one of about 1,100 evacuees in Houston who had to re-evacuate as Hurricane Rita became a category 5 and headed for the Texas coast. Houston officials said shelters might not hold up in a major hurricane and evacuees had to be moved to safer quarters. Many didn't want to move another time. Some felt they were being pursued by storms. Other evacuees were taken to Dallas, and some went to Louisiana, although that state was also being targeted by Hurricane Rita.

"Hell. It's been pure hell," said a mother of four. "I feel like a rag doll, people throwing me around." How long would it be before she was again in control of her own life and her children?

One young lady headed for Arkansas hoping it wasn't going to be like the Superdome where she slept on the ground for days. Her boyfriend was simply getting mad and impatient with authorities, but most of all with Mother Nature. Hollering at Mother Nature does no good so he hollered at the reporter and officials.

Problems Peculiar to New Orleans

Politicians have many more things than disasters to think about, of course. Before Hurricane Katrina, New Orleans Mayor Nagin was criticized for pleasing business interests and ignoring the poor. His predecessor, Marc Morial, left office in 2002 after serving two terms. Morial, who now heads the National Urban League, got credit for cleaning up the police department and creating a friendly climate for business. For some 30 years, blacks have mostly controlled city government with black majorities on the school board, the city council, a black police chief, and district attorney.

Tourism, casinos, black professionals and business interests have profited despite accusations of cronyism, patronage, and influence peddling. However, the city's poverty rate has grown to three times the national average, so the poor of New Orleans seem poised to blame their mayor, city council, and the federal government.

Mayor Morial's platform was to hire a better police chief, to end corruption in the New Orleans Police Department, and to reduce the crime rate, and he did those things. A federal investigation uncovered evidence that dozens of New Orleans police officers were participating in corruption. Morial was also proud that he improved police morale by obtaining a pay raise for them, although it followed a horrific number of murders and an increased homicide rate.

Morial also added minorities and women to the list of government contractors. He helped create almost $1 billion in public construction projects including projects to improve the schools and the criminal justice system. In 1996, Morial helped win voter approval of Harrah's Casino, riverboat casinos, and video poker in New Orleans, all of which generate substantial revenues for city government. However, he had to issue an executive order in September of 2000 when the casino hit the headlines for seeking financial and regulatory relief from the state, city, and its major creditors after only ten months of operation.

Morial couldn't get the City Charter changed so that he could seek a third term as mayor. However, he surprised many by becoming the first black mayor to win a majority of the white vote while running against a white opponent. His father (Dutch Morial) had been the first black mayor of New Orleans.

Mayor Nagin's View of Environmental Challenges

Ray Nagin, also black, came into office with many good ideas and much vision about the problems of New Orleans. In 2004, he wrote Senators Mary Landrieu and John Breaux asking the Senate to support the Climate Stewardship Act, sponsored by Senators John McCain and Joseph Lieberman. He described how New Orleans was listed by the International Panel on Climate Change (IPCC) as "North America's most vulnerable city to climate change due to its low elevation, land subsidence, continuing sea level rise, predictions by the Union of Concerned Scientists about more frequent and severe hurricanes associated with climate change, and widespread mosquito borne illnesses."

The IPCC had also concluded that if heat-trapping emissions were not significantly reduced, sea levels would rise 1 to 3 feet over the next 100 years. Rising sea levels, he wrote the senators, would "inundate low-lying cities like New Orleans and threaten survival as a city and the safety of our citizens."

Other effects of sea level rise might cause warmer oceans, changes in rainfall patterns, more intense flooding in the Gulf Coast, and would spoil water supplies by saltwater intrusions. This would lead to loss of fisheries and habitats, and negatively impact the seafood industry, culture, and tourism in New Orleans. He added that oil and gas pipelines would be exposed to spills causing damage which would further degrade wetland ecosystems and render transportation waterways unusable.

He pled that the Stewardship Act would require industries to control pollution emitted into the atmosphere and warned that delaying action was more costly. He concluded, "Global warming poses threats to public health, fish and wildlife resources, and our economy."

Nagin's Plans Were Disrupted by the Hurricanes

The problems even more important than pollution were to land on New Orleans' doorstep with Hurricanes Katrina and Rita. Much of the New Orleans tax base floated away as taxpayers lost their property, their jobs, their schools and businesses, and even their lives in some cases. Local and state politicians must next worry about whether they can survive with the loss of tax revenue. New Orleans, like many cities, owes debts that it will not be able to pay for lack of taxpayer money. It is not easy for a city to declare bankruptcy, and bond ratings can become so impaired loans and lines of credit are hard to arrange.

Mayor Nagin told many after the hurricane that the city was broke, although it expected to receive some FEMA money. Health care costs were expected to rise but many of those on Medicare and Medicaid were relocated, so their dollars will not be spent in the Gulf Coast.

Gov. Kathleen Blanco's office has been working on proposals to simplify and hasten government-guaranteed loans to businesses affected by the storm. Who will have the confidence to loan money to New Orleans? Who will believe they can pay their debts?

Politicians like the mayor and the governor have tended to blame others when they did not get the help they expected or requested. The governor learned later that when she asked for government help, she should have been more specific since she was the on-site person with knowledge of the conditions at the Superdome and other places in New Orleans. It was left to the White House and FEMA to come up with specifics in the first hours and days of the Katrina aftermath.

She was ready when the second hurricane came three weeks later. By the time Hurricane Rita hit land, Blanco said more than 90 per cent of residents in southwestern parishes had evacuated. For those who had not, she had advised them to find a safe place on the highest ground or in the highest building they could find.

Hurricane Rita Set Back Recovery Even Further

As Rita approached, authorities called off the search for bodies from Katrina.

Earlier during the Katrina disaster, authorities provided few transportation vehicles out of the area. But criticisms of the mayor were nothing when compared to criticisms of FEMA director Michael Brown. He reported early on that the rising death toll was due to those "who did not heed the advance warnings." Brown's lack of sensitivity and lack of knowledge about conditions came to the fore, and compounded his lack of effective management of the emergency. Reporters checked on his background and exposed his lack of experience and knowledge about emergency management.

Within days, he was replaced locally, and once returned to FEMA headquarters, resigned his post. FEMA is the agency that conducts many of the disaster scenarios and training of leaders. The head of the agency needs to be trained in taking command, even more than subordinates.

At the top of any organization, business, store, hospital, schools, and even homes, plans should be in place for communication, delegation, emergency supplies, and backup equipment in anticipation of unexpected problems. How many of us have carried out our responsibilities in those places to practice and prepare for emergencies? Officials such as the New Orleans mayor, the local Red Cross executive director, and the city council chairman have acknowledged the city's inability to pick up everyone who needed transportation to escape Katrina's floods.

Research by the University of New Orleans had recently predicted that as many as 60% of the southeast Louisiana parishes would stay at home even if warned of a Category 3 hurricane. About 36 hours before the storm hit, newspapers were writing that those who lacked transportation also lacked funds for hotel rooms and nobody was providing those funds.

Yes, there had been exercises and scenarios providing city, county and state leaders with preparation for disasters, and even for hurricanes. But how ready are leaders to take charge after perhaps only an annual scenario of a few hours, which will have many different components than a real event?

First responders like fire and police practice constantly for emergencies and their daily work places them in situations where they must carry out what they learn in an almost automatic fashion. Should leaders go through more and better scenarios to know how to take charge and summon help in disasters? Undoubtedly, they should. Scenarios in the future should include the details of Hurricanes Katrina and Rita to be most effective for future disasters.

Politicians were much readier for Hurricane Rita than for Hurricane Katrina. They were prepared for fires, floods, loss of electricity, and communications. They alerted citizens to leave much sooner. The cities of Texas were left mainly in the hands of first responders by the time Rita hit land. By that time, it had decreased in strength and the winds were down to a category 3 hurricane, unlike Katrina which remained a category 4 when it hit land.

The old historic area of Galveston (the Strand) suffered from fires as Hurricane Rita blew down electric wires. The Galveston fire chief likened the scene to a war zone with the live wires "shooting fire across the street." One of the buildings that caught fire in Galveston was one hundred years old. Several fires also occurred in and around Houston, Pasadena, and other towns. But the land was not as low as that in New Orleans, the residents were mainly gone, and emergency personnel could deal with technical problems and had little loss of life.

Potential for Big Damage from Hurricane Rita

The Texas coast from Corpus Christi, Texas, to Lake Charles, Louisiana, had some one-quarter of the nation's oil refineries and it was feared that they would be hit by Hurricane Rita, but they were spared. They had expected the consequences to be as severe as the loss of the city of New Orleans, but it didn't turn out to be as bad as feared. Will the survivors and evacuees say that officials cried "Wolf" when they didn't need to? Only the next disaster will show whether evacuation orders are heeded or ignored.

Damage to the vital concentration of oil refineries along the coast appeared relatively light. Valero Energy Corp. said its 255,000-barrel-per-day Port Arthur refinery sustained significant damage to two cooling towers and a flare stack and would need at least two weeks for repairs.

Rita roared ashore September 23rd close to the Texas-Louisiana border as a Category 3 hurricane with top winds of 120 mph and warnings of up to 25 inches of rain. By the next day, it was a tropical depression with top sustained winds of 20 mph located about 20 miles southeast of Hot Springs, Arkansas.

Again, some of the worst flooding occurred along the Louisiana coast, where floodwaters were nine feet deep near the town of Abbeville. In Cameron Parish, sheriff's deputies watched appliances and what appeared to be parts of homes swirling in the waters of the Intracoastal Waterway.

About 500 people were rescued from high waters south of New Orleans, some by helicopters. Another 15 to 25 people were reported stranded farther west along the shoreline of Vermilion Parish, but searches were postponed until September 25 because of high winds. Elsewhere, a portion of Interstate 10 over the Calcasieu River in Lake Charles was closed after barges broke loose from their moorings and slammed into the bridge.

New Orleans, devastated by Katrina barely three weeks earlier, endured a second straight day of new flooding that disrupted recovery plans. The Army Corps of Engineers said it would need at least two weeks to pump water from the most heavily flooded neighborhoods, the impoverished Lower Ninth Ward after crews plugged a series of levee breaches.

Texas officials planned for an orderly return of the nearly three million people who had fled ahead of the menacing storm, setting up regions that would reopen to evacuees after only two or three days despite electricity outages and local flooding and fires.

President Bush urged citizens not to prematurely assume the danger was over. "Even though the storm has passed the coastline, the situation is still dangerous because of potential flooding," he said. "People who are safe now ought to remain in safe conditions."

The return of residents on September 26th prompted federal officials, including President Bush, to voice concerns that the cities of Texas and Louisiana weren't yet ready to receive them.

Earlier, New Orleans Mayor Nagin defended his decision to let people back in. "The citizens of New Orleans deserve the opportunity to see what they have left and what they can salvage," Nagin told Fox News in response to warnings from the federal official in charge in New Orleans, Coast Guard Vice Adm. Thad Allen.

"I'm a little surprised the admiral came out publicly on this," he added. "Maybe since I've been away a day or two, maybe he's the new crowned federal mayor of New Orleans."

These disputes between local politicians and federal leaders who have written contracts with cities to take charge in emergencies suggest that protocols should be developed and reviewed in the event of disasters. This is why practices and exercises are so very necessary.

President George Bush said four days after Hurricane Rita that Congress ought to consider giving the U.S. military the lead role in responding to natural disasters. This idea was suggested by Maj. Gen. John White who described how five helicopters showed up at the same time to rescue one person in New Orleans after Hurricane Katrina hit. Bush said Congress would have to consider under what circumstance the Department of Defense should become the lead agency in coordinating and responding to a disaster.

"Clearly, in the case of a terrorist attack, that would be the case. But is there a natural disaster...of a certain size that would then enable the Defense Department to become the lead agency in coordinating and leading the response effort?" Bush asked.

Bush wanted to make sure there was a "very clear line of authority'" in the event of another major catastrophe, whether it is another storm like Katrina or an avian flu outbreak, and he appears to believe that the Department of Defense has that capability. The next step will be talks with congressional leaders.

Bush's recommendation to change the disaster and emergency responses to the military met with immediate negative reaction by mayors and governors across the country.

Louisiana Sen. Mary Landrieu expressed reticence about that approach on CNN's *Late Edition*. Landrieu said the military has a strong role to play "but so do our governors and our local elected officials....we do have a democracy and a citizenship that has elected mayors, county commissioners and governors particularly. I'm not sure the governors association or all the mayors in America would be willing to step aside," she said.

Turf battles continue to be argued over in the wake of Hurricanes Katrina and Rita. Egos may be hurt if people are not in charge of things. The citizens of the areas vulnerable to disasters are more likely to be truly hurt than politicians. Can these issues of who is in charge of what be worked out before the next major crisis arises? Perhaps the old Italian politician Machiavelli was right—ambition is so strong that people can never be satisfied because they can't all be in charge.

6

Press Probably Saved Lives

"The best newspapermen I know are those most thrilled by the daily pump of city room excitements; they long fondly for a 'good murder'; they pray that assassinations, wars, catastrophes break on their editions."

—Pete Hamill

Within three days of the hurricane, civil rights leader Jesse Jackson fiercely criticized the Bush administration over the devastation of Hurricane Katrina, charging that it could have been handled better, modified by better preparation, and claimed black people were being prevented from playing "top relief roles." The media was there to print what he said.

Politicians, including the President, use the media to look good. That's nothing new. The media is always the outlet for all kinds of presidential decisions, which usually have an acceptable political "spin" on them.

After Hurricane Katrina, a *Wall Street Journal* editorial cited a U.S. Army Corps of Engineers official in New Orleans who stated that although there was a disaster plan in place, the hurricane was stronger than expected. President Bush, it may be recalled, declared to the press that nobody "anticipated the breach of the levees."

The Bush administration has tended to *minimize* the long history of warnings to upgrade New Orleans levees and dangers weak levees might pose to the population. They do that minimizing by their comments and statements to the press who get that version of the truth out to the public.

Congress chose not to approve funding adequate to protect wetlands, maintain or enhance the levee system, and prepare for major evacuation, and the media was the outlet for their decisions as well. Depending on the experience of newspaper reporters, they may or may not realize how they are used to send information that is not completely neutral or objective.

New Orleans and the Mayor

New Orleans Mayor Nagin was a cable TV executive at Cox Communications who won election as mayor in 2002. A man of working class parents, baseball was a route to a scholarship at Tuskegee University where he earned an accounting degree. He then went to Detroit to work for General Motors, and ventured on to Los Angeles and Dallas to work for financial institutions. After he returned to work with Cox in New Orleans, he took a large pay cut for a mayoral salary of just over $100,000.

During the hurricane crisis, he was praised for maintaining calm and realistic attitudes while people clamored about him. Yes, he lost his temper and used cuss words and railed at federal officials for ignoring or delaying his request. However, he understood the power of the press and used his opportunity with the media to ask for help for his city saying, "This is a desperate SOS. Currently, the Convention Center is unsanitary and unsafe and we are running out of supplies for 15,000 or 20,000 people...For the next two or three months, in this area, there will not be any commerce at all. No electricity, no restaurants."

There are times when the media is used and ought to be used to summon help, deliver information, develop thinking, and influence people. But did careless reporters who may or may not be gullible let themselves go too far in delivering the news in New Orleans?

Mayor Nagin and Police Supervisor Eddie Compass apparently told reporters of rumors about conditions at the Superdome while it was being used as a shelter. They claimed that people were being killed, babies raped, and that citizens and police were involved in shoot-outs outside the stadium. Further investigation has found that these accusations were unfounded and yet the news media picked up the stories and shocked readers with the violence among victims. There was even a report by one media outlet claiming that blacks in New Orleans had resorted to cannibalism for survival. There was not truth to that rumor.

The media must be used carefully because it controls public opinion in America today. In fact, the media's opinion has shaped public opinion for the last several decades. Research by the Statistics Department in Washington, D.C., showed that at the end of the 1990s, the average American watched about 1,000 hours of network television annually, 400 hours of cable television, read the newspaper for 150 hours and spent about 100 hours reading magazines.

Obviously, the media is no different from any other business in a capitalistic society like America and is out to make a profit. Since the goal of the press is to sell their work, they often select news to entertain or titillate the senses rather than providing objective information to readers. Most media executives believe that conflict sells. If there is no conflict, there is no news. They have a saying that "If it bleeds, it leads!" Often the media attempts to create conflict where there is none to sell their news. To do this, they may ask provocative questions intended to stir up emotions and make a more interesting story.

On a daily basis, reporters need material for stories and all they have to offer in return is media exposure and sometimes flattery. Those who seek attention may be happy with exposure and flattering or even non-flattering coverage, as long as they get the coverage and attention.

The cynical handling of some issues by the media and their preference to depict conflict make society's problems harder to solve than they would otherwise be. Often, the press gets in the way of society solving its own problems. In fact, their sensationalism often incorrectly exaggerates menaces, making citizens more fearful than they need to be.

Reporters often irresponsibly convince the public by their slanted articles that public officials cannot be trusted. They do this by claiming to look into issues and calling it "investigative reporting." These "investigative reports" often yield alarming results that must be disclosed by reporters, or so they suggest, because they are being *hidden* from the public. The media then proceed to bring out the worst in some persons in public life, driving serious candidates away and rewarding gutter fights. Our so-called "news" is the result of hundreds of judgment calls made by reporters and publishers instead of objective news reporting.

It is clear, also, that campaign media consultants can make or break presidential candidates. Television's interpretation of primary election results can influence voters, as media professionals tell the listening audience what they just heard, adding their own spin on things.

Speeches Aren't For Us, They're for the Media

Speeches are composed for the media even more than they are for the audience. Many more people will read speeches, commentaries on the speeches, and capsule comments the next day rather than listen to it themselves. Therefore, short pithy comments, speeches attacking opponents, and exciting phrases present reporters with juicer material than longer sentences containing complex information. There is no doubt that the trend of attacking the opponent and negative campaigning suits the media and their wish to sell stories.

The press claims to have the license to criticize and defame but too often they shun the license to be responsible in serving the public. They tend to serve the selfish side of the human spirit more than the striving, inquiring side. They should help voters and readers sift through facts, information, and propaganda, equipping them to evaluate issues intelligently. The press could ask the readers what issues they want candidates to discuss instead of putting their own questions to candidates. When they plan to interview public authorities, they could ask readers to submit questions for the interview, or they could examine the context in which they are going to delve. But instead of more serious research, they often simply create questions themselves.

It is true that speeches, wars, disasters, and violence must be reported. How it is presented determines whether the reader feels happy, angry, powerless, betrayed or involved with their institutions and political system. The focus of television is image, 30-second sound bite phrases, and only incidentally is upon the definition of real issues.

Investigation, explanation, objectivity, and fair-mindedness should be the tools of reporters. If they are not used, the media gets in the way of society solving its own problems. Irresponsible reporters contribute to the public's anger and distrust of their own public officials. The media's real goal should be to make what is important also interesting enough to learn about, understand, and use to improve life.

Sometimes members of the press recognize and write about these problems. Such a person was Kathleen Parker of *Tribune Media Ser-*

vices. She said in an article in September, 2003, "It is indeed too soon to pass judgment on Iraq, but bad news is what compels and sells. Journalism's once heroic goal of seeking truth has been subjugated, it seems, to the more commercially expedient mandate of 'sexing up' the news." It is refreshing that a prominent reporter like Kathleen Parker would point out the problems of her own profession.

Media Wants to Make Money—Surprise, Surprise!

The news media tend to cater and pay more curious attention to the abnormal than the normal. Just as curiosity killed the cat, such media philosophy helps to kill a city.

Our TV media news programs now have 17 minutes of news in a 30-minute segment. There is a general lack of community news and important information, in favor of "smart shoppers" plugging products and "investigative journalism" with reporters trying to create sensationalism through exposing bad business practices.

Controversies abound. Dan Rather's *60 Minutes* story on President Bush's National Guard service during the Vietnam War ended Rather's career, as the media rushed forward with the supposed results of "investigative journalism."

Network news is slipping. People have deserted to cable news, websites, videogames and whatever. Thus the major news programs and anchors try to bring them back with dramatic news stories.

Extremists get a disproportionate amount of attention in our media, and therefore create a false sense of alienation between those in the mainstream of life. The media often give people a chance to vent their spleen. Good news gets the back page because controversy is what sells.

Today, there is dissension among the media as they fight for their share of the audience. When Fox News overwhelmed CNN during the Iraq War, Ted Turner warned his competitors that just because ratings were higher, they weren't necessarily better.

Advertisements cater to basic appetites, as do news stories. Sexual material sells well and stories about sex sell well. Big disturbances and loud noises catch the headlines, the camera, and the microphones. People who are quiet, reflective, and weigh all the issues and act with moderation are toast in the papers.

The press should be fair, even-handed, and give equal space to all sides of issues. Consumers of media, you and I, have a special impact on a city's image and therefore a special responsibility. They and we should balance negatives with positives to keep a city's image in balance. Improper or erroneous reporting can wreck a city.

But now let us talk about stories that were helpful to the public. An article entitled "Left Behind" was published in the *Times-Picayune* in 2002. The New Orleans newspaper warned, "Once it's certain a major storm is about to hit, evacuation offers the best chance for survival…And 100,000 people without transportation will be especially threatened…A large population of low-income residents do not own cars and would have to depend on an untested emergency public transportation system."

This article could have caught the attention of poor people who might not have transportation to evacuate in the event of a major storm. Reporters had noticed a lack of preparation that government leaders were ignoring. The press acted responsibly in reporting this anxiety-provoking article.

A little over a month before Hurricane Katrina hit, a July 24 article in the *Times-Picayune* reported, "City, state and federal emergency officials are preparing to give the poorest of New Orleans' poor a historically blunt message: in the event of a major hurricane, you're on your own." It was the same message, and it was a needed message. It turned out to be true, even though it appeared to be one more article criticizing those in power.

The media should be an objective force, helping to inform citizens about news and what is going on that might affect them. It should give them information that will enable them to act and deal with issues responsibly. It should never manufacture problems, or exaggerate them, or confuse them. Journalists may have their own views but these should not be displayed in news articles. Columnists may, of course, speak for themselves, and that is what they are paid for. Their opinions are requested but that is not true of those who are to report objectively about news.

Reporters were at their best during most of the Katrina and Rita Hurricane news coverage. Many risked their own lives to be in horrendous weather with little backup and available help if they were to run into life-threatening situations. This nation and others were tuned in to see the damage wrought by Hurricane Katrina. The media delivered stories from every possible angle and most of the on-site reporters delivered it objectively.

Columnists were a problem. Some pontificated about whether global warming had anything to do with the more severe and frequent hurricanes and, as if they were meteorologists, concluded that it did or did not.

The rumors about human behavior in and around the Superdome did not serve New Orleans well. It tainted the opinion of people throughout the world about the real nature of people during a crisis and the real nature of the victims. For this, they must bear some blame and re-examine their purpose and their power.

The best thing the media did for those in the path of Hurricane Rita was to cover Katrina's wrath so people could decide whether to evacuate. The media pictures and the coverage of drownings and devastation were undoubtedly why an epic three million people evacuated before Hurricane Rita, saving countless lives.

7

Pressure Groups Created Pre-disaster Problems

○ ○
"Everyone sees his own cause as just."

—Anonymous

The New Orleans *Times-Picayune* reported on November 18, 1998, that St. Bernard Parish city fathers, just east of New Orleans, asked Congress for money to study the New Orleans hurricane levee system. That request followed a plea from the U.S. Army Corps of Engineers *suggesting that local governments lobby federal officials* to upgrade the levee system to withstand a Category 4 or 5 hurricane. The Corps was operating as a pressure group, and thanks to their pressure, flood prevention issues were brought before Congress.

In 1990, Congress created a task force to establish a greater buffer of wetlands to protect New Orleans. When funding was insufficient, the project was able only to minimally slow down the destruction of the wetlands. The U.S. Army Corps of Engineers officials have told reporters that they believe the war in Iraq and homeland security reorganization were the main reasons that Congress reduced funds to New Orleans flood-control projects.

Civil Rights groups constitute some of the most vocal pressure groups. Jesse Jackson criticized President George W. Bush to the press over Hurricane Katrina claiming that blacks were being kept out of top relief efforts. He pointed out that many trapped in New Orleans were not only poor but also black. He said, "There is a historical indifference to the pain of poor people, and black people."

Jackson led a bus convoy to New Orleans to rescue students trapped at Xavier University, and said he was shocked by sights of refugees trying to flee the city. He claimed that relief efforts were unacceptable. He gave statistics about the poor in New Orleans, stating that they were poor, black, old, sick, and many had no private transportation. He criticized calling in former presidents for fund raising and leaving blacks out of such efforts.

Nation of Islam leader Louis Farrakhan told followers in Charlotte, North Carolina, a week after Hurricane Katrina hit New Orleans that levees were deliberately destroyed to flood blacks. The influential preacher said, "I heard from a very reliable source who saw a 25-foot deep crater under the levee breach...It may have been blown up to destroy the black part of town and keep the white part dry." He didn't say who he thought blew up the levees.

The Muslim minister followed up by going to Houston where many evacuees were taken. There he listened to survivors' stories such as one lady who said that she and her sister waved and hollered for helicopters to rescue them after the hurricane, but were ignored. Farrakhan responded that "FEMA is too white to represent us and so is the Red Cross, so we're going to demand our place at the table." He said that he and the Millions More Movement—a group of black leaders organizing the 10th anniversary of the Million Man March—would do everything in its power "to ease the suffering of our brothers and sisters."

Such pressure groups create much unrest, distrust, and anger when they spread rumors such as this. Farrakhan has received much critical press for his comments and attempts to stir up racial animosity.

Business Pressure Groups

Some pressure groups represent a section of the public on some particular issue. A group may become popular when people feel that the government doesn't listen to their point of view. Other groups pressure local government to undertake projects that would benefit them financially. A good example in Detroit was Super Bowl XL. Backers maintained that the big game would generate $302 million. It was predicted that 125,000 out-of-towners would come to Michigan in the dead of winter to spend an average of $2,500 each. The real beneficiaries, however, were the National Football League, the stadium owners and big businesses. The losers are the taxpayers and local businesses. Imagine if the $302 million was invested directly in schools, the police, and local businesses?

One pressure group in New Orleans was the casino industry. As casino architect Robert DiLeonardo said, "My job is to create an environment that relaxes morality." The association between casinos, gambling, flesh shops, bars, Mardi Gras, overt sexuality, and "the Big Easy" just seemed natural, so gambling came into New Orleans with little resistance by the residents.

In fact, it has spurred Louisiana to open several residential rehabilitation programs for compulsive gamblers, and advocates Gamblers Anonymous meetings that occur weekly in New Orleans. A 2002 study of gambling by the Louisiana Office for Addictive Disorders estimated that there could be 50,000 problem gamblers in the state. There is a statewide gambler's "helpline" which received almost 50,000 calls in the last fiscal year that ended in 2004.

Gambling in Louisiana just began in the last decade, with the presumed advantage of offering jobs. Potential jobs were estimated at between 10,000 and 25,000, but the four casinos have not been able to fulfill those expectations. The Fair Grounds Race Track recently declared bankruptcy and some say that was because it could not compete for gambling dollars with local casinos. Bally's has been unable to compete with Harrah's New Orleans Casino and the casinos were reportedly considering a merger in the year before Katrina.

All together, the four area casinos have brought more than $100 million to the state treasury in the last year, as well as paying state taxes and other fees charged by local jurisdictions. They have brought problems, however, to some politicians such as former governor Edwin Edwards who was convicted on charges connected to licensing casinos.

As casinos continued to expand, they were in the process of developing a 450-room hotel which was to open in 2006. Harrah's claimed to draw customers from at least a 350-mile radius of New Orleans, and was intended to become the largest tourist attraction in Louisiana. In 2005, Harrah's sweetened the pot for the New Orleans City Council by offering community development grants up to $50,000 per business up to a total of $250,000, to help small businesses in the New Orleans area, as long as the businesses promised to promote gambling and Harrah's. The status of the casino expansion is questionable in view of the economic turnaround following the hurricanes.

Natural Resource Pressure Groups

Meanwhile oystermen and shrimpers make up yet another pressure group. They have complained that the proposals from both the scientists and the engineers to make some changes in the water composition as they restructure levees would ruin their fishing grounds, their income, and the food supply from seafood.

Oil and gas interests make up yet another pressure group. Robert J. Samuelson's *Newsweek* article September 19 about our vulnerability to a major cutoff of oil in New Orleans clarifies the importance of Louisiana's oil and gas producers. More than 60% of our oil is for transportation, mainly road travel, he says. Most of our vehicles are SUVs and high gas consumers, and we are likely to increase the number of vehicles in the U.S. by at least 50% over the next generation due to population increase and higher incomes. Domestic U.S. oil and gas can cushion disruptions in the supply from the Persian Gulf and Russia. But just as foreign oil sources could be disturbed by war, terrorism or other causes, so can our own supply be hurt by hurricane damage to refineries in Louisiana and Texas.

As long as we continue our love affair with the automobile, oil and gas pressures will be felt in states that are major fuel suppliers. They are, like other pressure groups, going to demand special protection from city, county, state, and especially national authorities. They will have much influence over whether New Orleans will be rebuilt and whether flood controls will be sufficient to protect their interests.

In the weeks following Hurricanes Katrina and Rita, oil production was down to 75% because of storm damage. The President's advice to gas guzzling American drivers was to ask them not to make unnecessary automobile trips.

8

People Are Only Human

○ ○

"It is more useful to watch a man in times of peril, and in adversity to discern what kind of man he is; for when at last words of truth are drawn from the depths of his heart, and the mask is torn off, reality remains."

—*Lucretius.*

In disasters, people are forced to focus their attention on the current problem and make decisions which can be life-saving. They think more clearly and seem to be very realistic in such crises. Naturally, one of the first reactions of people involved in hurricanes, burning buildings, or other disasters is disbelief. That is usually when someone says, "Let's get out of here!" That kind of exclamation makes people flee, in case they had any doubt about their danger.

Some people move quickly and some slowly. Some help others and delay their own exit. Those facing death are usually kinder to one another than normal. They panic only rarely. Even though events can be loud and noisy, we put them to the back of our minds because we have lived in relative safety throughout our lives and cannot believe that we are in danger. Our mind requires several seconds to handle each new piece of information. For example, we see this mental processing in television replays of President Bush in a classroom on 9/11 as he received a report about the first tower hit by terrorists. Minutes later a report about the second tower (therefore no accidental airplane crash) is given to him and we can sympathize with his slow reaction as he tried to understand the implications of the information and decide what to do.

Of course, more information and more stress slows down processing and decision-making. We sometimes deny or ignore facts because they require us to figure out something different to do. So, when we are asked to evacuate an area, we check several sources—friends, family, television, radio, and others before deciding what to do.

On the other hand, if you don't know what to do and someone says "follow me," it may save your life rather than doing nothing. There is some notion that disasters turn us into animals, driven by instinct. Researchers find that on the contrary, rather than becoming panicked, people do nothing. That freezing behavior (fight, flight or fright are first reactions to crises) may be good for animals that play dead when attacked. They do this because animals that eat struggling prey assume it is healthy whereas prey that doesn't struggle may be unhealthy.

People don't really go nuts or panic or run wildly as much as movies suggest. They actually become very helpful to each other in most situations. They take care of themselves surprisingly well until help arrives, figuring out what they need to do and how to do it with little, if any, prior training. Of course, someone with some prior emergency training often becomes a temporary leader among a group of people in a disaster. Others look to them for direction so that they can save each other and save themselves.

These patterns of fairly orderly and altruistic human behavior can be seen in hurricanes, volcanic eruptions, earthquakes, tornadoes, floods, and all sorts of natural calamities. The terrorist attack of September 11, 2001, showed that same behavior was present even though the attack was not an "act of God." Even those aboard United Airlines Flight 93 were able to find some way to work together and overcome a worse disaster by preventing terrorists from yet another attack. The help people offered to each other in crises like 9/11 and in natural disasters is truly inspirational.

Our society's economic basis may be competition and capitalism—selling goods and services for the most money—but disasters show that the bottom line is our mutual care and humanitarianism. During the hurricanes, not only were people helping each other face to face, but those outside the danger zones offered money, homes, schools, jobs, transportation, medical and physical care, and all sorts of help to those in need.

What About the Bad Guys?

Despite a good deal of basic humanitarianism, we wonder why immorality began to take place in some areas so quickly. There was some looting and lawlessness within the first days after Hurricane Katrina. Some residents took food, diapers and necessities from abandoned groceries but others carried off television sets, alcohol, and weapons. Yes, some New Orleans citizens quickly resorted to uncivilized behavior. Law enforcement officials could not be everywhere and were often not around to stop those few bad people who set a fire, shot at rescuers, and hijacked a car. But they were not the norm; they were the exception and that is why they got the attention of the media.

Each of us has a different way to handle the stress and trauma of a disaster. Those coping reactions can be positive or negative, depending on whether selfishness or selflessness has become a part of the personality. A self-centered person will be more concerned about his own survival and safety than with the survival and safety of others. A selfless person will be lending his help and common sense to others in emergencies.

The longer stress goes on, the harder it is to maintain positive coping reactions, of course. The Superdome is a good example of a place that was first a welcome shelter from the storm and flood but became more like a concentration camp as people ran out of supplies and no rescue came soon. Some National Guard troops and local law enforcement officers could not maintain order and could not appease angry victims who ran short of food, water, and facilities as they waited for authorities to come to their rescue. It now appears that news stories exaggerated the extent of lawlessness, however, because so many survivors have good stories of mutual aid to tell. In those desperate conditions, some people sank to a lower level of morals than others.

The basic needs for physical survival, security, and desire for control over their lives kick in when people are in desperate situations. Some fell into basic reactions of distrust and violence, which they would have learned through a lifetime of being unprotected by family. That could have been compounded if they had past experiences with indifferent social service agencies. Under stress, they might feel they must defend themselves from others who approached them at the Superdome or on flooded streets, even though the others might have been prepared to offer protective services.

Others were more optimistic, because they must have had a lifetime of good experiences at the hands of others, and felt they were going to be saved. They chose to take only what they needed until rescue occurred.

Still others resorted to a childlike behavior and imitated the uncivilized infant who wants what he wants when he wants it and cannot brook delays or waiting. In other words, some began to loot and use other people and other property to enrich themselves.

President Bush told Diane Sawyer on television that there should be a no-tolerance approach to looters. "I think there ought to be zero tolerance of people breaking the law during an emergency such as this, whether it be looting or price gouging at the gasoline pump, or taking advantage of charitable giving or insurance fraud," Bush said. But many believed that those who took food for their families should not be blamed. Television viewers were less forgiving to looters who took items to sell or trade such as television sets after Katrina.

Depending on their morals and upbringing, some people take things and others don't. Tragedies cause one's true character to emerge. And that includes those across the nation and elsewhere who have generously donated millions for survivors, and even opened their homes to Katrina victims.

Some who are more giving had even gone to the trouble to be trained to help in emergencies, like those who undergo Red Cross CPR training. The Community Emergency Response Team (CERT) volunteers are regular citizens who have been trained to help communities survive for the first two days before professional services may be available. Many CERT volunteers arrived to assist in the documentation and check-in of survivors who arrived at the Superdome. Volunteers assisted citizens to get from the busses to the check-in locations.

What Makes People Commit Crime?

I think all good folks know what makes people commit crime. In fact, I guess we've all known it for hundreds and even thousands of years.

Some 2,500 years ago, Plato wrote about good and bad behavior. He described Socrates discussing that subject with others in Athens,

Greece. He explained that people make decisions based on seeking pleasure and avoiding pain. But he said that some people are overcome by pleasure or the quest for pleasure and do what they know is wrong. To make better decisions and be more virtuous, they must learn more, he said in *Protagoras* and *Meno*.

Plato and Socrates thought that virtue could be learned and better decisions could be made. All those who do therapy with criminals share that philosophy today or they wouldn't be trying to help. The key to living a good and virtuous life, according to Socrates was to "Know thyself."

John Stuart Mill described hedonism in such a way that criminals could easily be called hedonists. He said that hedonism was "the theory that a person always acts in such a way as to seek pleasure and avoid pain." He and his mentor, Jeremy Bentham, reasoned that when making choices about behavior, one should try to obtain the greater good for the greater number of people.

The words and philosophy of the Persian Omar Khayyam (who died in 1123) were translated by Edward Fitzgerald. Khayyam understood that some people regret their actions, apologize, and wish they could take it all back but it is too late once the act is done.

> The moving finger writes; and having writ,
> Moves on: nor all your piety nor wit
> Shall lure it back to cancel half a line,
> Nor all your tears wash out a word of it.

William Shakespeare undoubtedly understood why people choose to do criminal acts. He knew that if we were honest within, we would also be honest with others. In *Hamlet,* he wrote,

> This above all: To thine ownself be true,
> And it must follow, as the night the day,
> Thou canst not then be false to any man.

Thomas Huxley, like Socrates, thought that learning could overcome crime. "The only medicine for suffering, crime and all the other woes of mankind, is wisdom." But, of course, some people may learn and copy the bad behavior of parents, friends, and role models. If they had new and better models and people to imitate, would they likely make better choices in their behavior?

Sigmund Freud discussed the pleasure principle, and described how people are born with the goal: "I want what I want when I want it!" He called that selfish part of us the "id." As parents teach children the difference between right and wrong, a child accumulates "learnings" that lead to the development of a conscience. Freud called those "learnings" from parents the "superego." Once children become adults, they make choices about whether to follow their "id" or their "superego." Freud called these choices the "ego," saying that individuals develop "ego strength" to resist temptations.

Moral theoreticians came up with stages that children go through in the development of a conscience. Lawrence Kohlberg and Jane Loevinger and others say that in the beginning, all children are hedonists. Next, they learn to do things to avoid punishment and pain from parents and adults, and instead do things that bring love and pleasure. As they begin to interact with other older children, they learn to share, give and take, and follow rules so they can maintain friendships and avoid embarrassment and penalties. Next comes a realization that law and order in their community, school, workplace and society requires that everyone follow rules or they will run into trouble with authorities. Most youth finally accept the social contract to "do unto others as you would have them do unto you."

Eventually, some people may choose to violate the social contract and suffer the consequences for what they think is a higher good. Americans who risked their lives to fight for independence from the British during the American Revolution would fall into this category.

Those who have reached at least the rule-following stage of moral development are less likely to return to their earlier selfish outlook. Those who, for a variety of reasons, do not develop a conscience, tend to be repeat offenders or recidivists. Police have long known that a large proportion of crimes are committed by a small number of criminals. If they are taken out of circulation, the crime rate is reduced.

Police Presence Can Deter Crime

There are other influences besides morals that persuade people to make criminal choices. Lawrence Cohen and Marcus Felson believe that crime occurs depending on conditions where people are, such as the likelihood of getting caught in a particular area. For example, Ronald Clarke and Derek Cornish (in the textbook *Criminology*) suggest that a would-be burglar examines a house thusly. If the house has a less visible entryway, if there are few who could see the criminal, if there is no visible alarm, if there is no barking dog, he is more likely to run the risk. He therefore chooses to avoid pain and gain pleasure/profit. But with no police around in New Orleans, gain exceeded pain.

Criminals begin early to choose how they will spend time, who they will spend it with, what they will do for pleasure, what they will use as excuses, who they will choose as victims, and what they will tolerate in the way of pain or punishment. There are no experts today who think, like some in the days of old, that criminals are born. They are made, as F.B.I. profiler John Douglas believes and expresses in books such as *The Anatomy of a Motive, Crime Classification Manual,* and his newest, *Anyone You Want Me to Be.*

In his book *The Anatomy of a Motive,* he writes:

> With the exception of a very few truly insane (and generally delusional) individuals, these men choose to do what they do....They are not compelled. They choose to do it because it makes them feel good.

Stanton Samenow, Ph.D., author of *Inside the Criminal Mind* and *The Criminal Personality,* believes the same thing. He wrote a book to warn parents about children's early behavior called *Before It's Too Late.* Samenow and his colleague Samuel Yochelson found in interviewing criminals that they had several "criminal patterns of thinking."

These patterns include the failure to put oneself in another's position, failure to consider injury to others, failure to assume obligation, the victim stance that one deserves more because of certain circumstances, the failure to see property as belonging to others, lack of interest in responsible performance, failure to endure adversity, and the power thrust to do something to become a "somebody."

John Douglas found some of the same thinking patterns also. In *The Anatomy of a Motive*, he wrote:

> Most violent offenders, we found after some study, had two factors warring within them. One was a feeling of superiority, grandiosity: social mores were not meant for them; they were too smart or too clever to have to start at the bottom and work their way up, or to live by the normal rules that govern a relationship. The other, equally strong feeling was of inadequacy, of not being able to measure up, of knowing they were losers no matter what they did.

Despite those feelings, Douglas emphasized how criminals make choices about what they are going to do. Just as the criminal considers his options, so do potential victims. People choose where they will go and when (dark is the criminal's friend), whether there are risks in the environment, how much they care to protect their living space such as their house or car, and how prepared they are physically (strength, self-defense training, etc.) and with safety items (pepper spray, cell phone, weapon, siren, etc.) to encounter a criminal.

There are more things that influence a criminal as he decides whether to risk getting caught for the pleasure and profit he might gain. If the community looks disorderly, he may suspect that residents are less vigilant and police presence is less responsive.

James Q. Wilson and George Kelling introduced this "broken window" theory in 1982. Their article suggested that attention must be paid to the little things. For example, a broken window that is not repaired tells the residents and the criminals that nobody is in charge. As residents go inside for fear of an uncontrolled neighborhood, nobody is present to see the criminals. Social order and civility are restored when residents and police care for the neighborhood.

This is something that will be sorely tested as New Orleans rebuilds itself. Will the police be able to randomly cruise neighborhoods making their presence felt to reassure citizens and deter criminals? And will good people and good neighbors prevail and help their fellow man rebuild communities when there is less stress and less need to unite?

What Psychological Effects Will Katrina Have on Victims?

Hurricane Katrina has changed the lives of its survivors, rescue workers and even those who merely watched the catastrophe on television and read about it. Many survivors, especially those who are older, will suffer from the loss of loved ones, property, money, possessions, jobs, rank, savings, illness, injury, and a variety of losses. Losses make people angry, depressed, and scared. Sometimes those who have lost much tend to withdraw from others and from life to avoid further hurt.

The younger people are, the less they have lost in such a disaster. Even those who have lost a lot can be extremely resilient, especially if the losses are only short-term and not lengthy such as being in a POW camp for years.

One of the greatest stressors is the unknown. Even when survivors are safe, their worry about the whereabouts and the fate of family members and friends still takes a heavy toll. The hurricane victims had to worry about whether they have houses to return to. They worry about whether they have a job to return to, a salary to receive, friends to continue seeing, and even familiar schools and churches to attend.

They worry about whether social services and insurance will be adequate to help them reclaim a decent life. These worries can distract them from other thoughts, leave them tense and nervous, unable to concentrate or remember accurately, and intellectually drained at the very time when they must make decisions about their daily welfare.

The loss of control over their lives and the need to depend on others is a sudden reversal of life, as if an adult were forced to return to childhood where others made decisions for him. Blanche DuBois in Tennessee Williams' *Streetcar Named Desire* disliked having "to depend on the kindness of strangers." So do survivors who must wait for others to make decisions for them. That role reversal is frustrating and rage-making.

In New Orleans, which had just drained nearly all the flood water from Katrina, Rita's wind and rain quickly broke through a patched levee into the already wasted Lower Ninth Ward and parts of nearby St. Bernard Parish. The water rose several feet high. Evacuees from the devastated area were in despair. "It's like looking at a murder," said one man. "The first time is bad. After that, you numb up."

Many of the victims have seen horrible things, things that are usually seen by no one other than police and firefighters. These sights bring shock, fear, disgust, mourning, and despair to those unprepared for them. Those experiences sometimes leave victims with nightmares, physical reactions like headaches, nausea, and weight loss. Some survivors, though probably not many, could develop post-traumatic stress disorder for a few weeks. The longer stress lasts, the more likely it is to disrupt lives.

Talking with others, including social service representatives and professional counselors, will help most people sort out what they've been through. After they have come to terms with the reactions they had, it is easier to move on from this point in their lives. It will be important for survivors to be debriefed, talk it out, process it, and defuse their experiences and reactions. They will want to get back into a normal lifestyle in which they regain control over their lives. Returning children to school, getting back in touch with loved ones and friends, getting jobs and drawing salaries will bring calm to the chaos they experienced.

Not only victims suffer. So do rescue workers. Two New Orleans policemen killed themselves. One, Sgt. Paul Accardo, became severely depressed when he lost control of how to help so many who seemed to need him. He was overwhelmed and suffered from the same symptoms described above—inability to focus and concentrate, difficulty remembering, and loss of normal reasoning. Quite often, those who commit suicide lose track of time and believe that things will be as bad forever as they are at this moment. Unable to keep the perspective that things will eventually change, they want to stop the pain and helplessness that they feel, without waiting for change to come. They are choosing a permanent solution to a temporary problem.

It was expected that people who saw the devastation of one hurricane (Katrina) would be readier to heed warnings for the next hurricane (Rita) three weeks later. Indeed, the majority of people in Texas and Louisiana did heed the warnings to evacuate. At least 2.8 million people fled a nearly 500-mile stretch of the Louisiana-Texas coastline in an evacuation that caused monumental traffic jams in which hundreds of cars broke down or ran out of gas. Traffic was still bumper-to-bumper the night of Hurricane Rita's landfall from the outskirts of Houston toward Austin and Dallas.

A Rice University student said it took about 18 hours to get to the interstate, plus another six hours to get to San Antonio, trips that would have taken only minutes at other times. Some drivers conserved fuel during their trip by pushing their car and shifting into neutral whenever possible. It was nerve-wracking for drivers and passengers to be in that kind of a situation for so many hours.

Downtown Beaumont was deserted, and buildings were boarded up. About 90 per cent of Galveston's residents left but one lady stayed and suffered burns when the building she was in caught fire. Another said she stayed because she would rather die in her house than on the street.

The rescue workers in Texas during Hurricane Rita were different from those in New Orleans during Hurricane Katrina. Usually trained rescue workers are very resilient, but in Katrina, these rescuers themselves had gone through the same losses and uncertainties as the victims they were trying to help. They had to weigh whom to deal with immediately, the victims in front of them, the looters and evil-doers, or their own loved ones who may be waiting for them somewhere in need.

Even the public who were not directly involved in events underwent some trauma. Those who watched the scenes on television from their living room saw shocking and depressing images. It is hard to realize that uncontrollable forces in nature could upset anyone anywhere. Children are always more vulnerable to such images and can develop fears of things that suggest they could be hurt or left unprotected. Parental explanation and reassurance is very valuable to them.

Children look to their parents for reassurance and as role models on how to behave. Parents who are very stressed out and anxious can make children worried and anxious. As they try to manage difficult situations after disasters, parents can be short-tempered and impatient with their children. Of course, young children can be more irritable and even uncontrollable.

It's difficult to do but the best thing parents or caregivers can do is resume some sort of routine. Even in shelters, routines can be established. For example, they can study, read, play, eat, or have bedtime stories at a certain time. Children must know that they still have to obey rules. It is reassuring for them to know that adults are in charge and will take care of things. Children don't want to believe that they are in charge or that nobody is in charge. Parents can help by letting children know that parents love them, and will keep the family together and safe.

With hundreds of thousands of children forced from home by Hurricane Katrina, the U.S. Secretary of Education Margaret Spellings worked out a plan to get them educated wherever they might be. She told reporters that as many as 330,000 students had to flee the Gulf Coast. She and others arranged for authorities to waive certain federal statutes to ease pressure on the states regarding school enrollment. Some of Hurricane Katrina's young survivors resumed some normal activity within days after leaving the Gulf Coast and were absorbed into many school districts across the nation. With an attitude of not wanting to shortchange children, receptive school systems arranged busing, school supplies, and free or reduced-price lunches.

Studies of Disaster Victims

Psychologists, who studied the flood victims of Johnstown, Pennsylvania, after the flood of 1985, found that survivors do not go to psychologists or other professionals for help. Professionals have to go to them. When survivors do seek out emotional help, they usually turn to their relatives, neighbors, friends, or clergy. In some cases, that may not be enough.

One main reaction that has shown up in flood victims is the "helplessness complex." Because others had to make decisions for survivors during a crisis, some survivors come to believe that this should continue and that they are not capable of making decisions for themselves. To a certain extent, citizens of New Orleans felt helpless to protect themselves from storms and floods, and expected their city council, mayor, police, fire, and other emergency services to make plans that would protect them. Will they become even more helpless as they accept services to be moved to other cities, financial aid, jobs, schools? What else will they feel they deserve, and at what point will they feel they can be responsible for their own lives and decisions again?

Why do some people react more negatively to disasters than others do? To understand this, psychologists have explored the issue of whether a person feels they have control over their life or are helpless. They found that if people believe they have more control over their own lives, they believe they can improve their situation. They feel less dependent on others to rescue them and expect more of themselves, so they recover more quickly.

Sometimes it is hard for survivors to believe in a benevolent world after a disaster destroys their lives and part of the world around them. They may search for the meaning of life after they have been buffeted about by "acts of God." In other words, their faith may be shaken.

In summary, poverty levels, general economic conditions, perceived control, and previous experience all seem to have effects on responses and recovery and psychosocial adjustments from flooding disasters. These and other things need to be considered when planning for post-disaster recovery.

The Special Case of Children

The responses of children to disasters have been studied following natural disasters. Results show how important it is to interview children directly. Children reported more anxiety symptoms than parents reported for their children. Children as young as six years of age reported emotional problems of which the parents appeared unaware. Children most likely to be adversely affected were those with a pre-existing disorder and those with parents who also reported a high number of anxiety or depression symptoms in themselves.

What about those just watching disasters on television or just reading about them in newspapers? The past century has been characterized by changes in technology, politics, biology, behavior, and the brain, and we can study people inside and out now. Additionally, these days we see and hear more news of natural disasters, catastrophes, and genocides on radio, television, and newspapers. Does seeing all these negative events and bad news affect us? In the past, traumatic stress involved only victims. Now it can involve those who watch the news. The most important finding was that it bothers people more if they can't do anything about what they see.

Adults, unlike children, can do something about the hazards depicted in news stories of calamities. Adults can participate in rescue efforts, in donations of goods and money for victims, and in preparations for their own protection.

Children can do nothing and therefore feel helpless, so adults must make plans for them. The days after the hurricanes contained many news stories of children raising money for victims with a lemonade stand or some other projects. This gave children a sense that they could do something about what they saw—they weren't helpless.

Some adults decided to improve their own protection, after seeing the disasters. That protection might take the form of training for neighborhoods to be prepared for emergencies until professional responders can reach them. It could be that fire and police plans for responding to emergencies will be beefed up. Companies, schools, churches, homes, care centers for young and old may review and improve their plans for coping with catastrophes.

The result will be less dependence upon others to ensure our safety and more proactive efforts to set up even better plans for surviving emergencies. The fact that adults can do something about what they see helps them recover, whereas those who feel helpless to do anything suffer more.

Arizona's Governor Janet Napolitano, like many other governors, called for action and review within weeks of Hurricane Katrina. In an *Arizona Republic* article on September 14, 2005, by Chip Scutari, the governor referred to the disaster as "a gigantic wake-up call." She directed her disaster response team to revise, modernize, condense, and simplify the state's emergency plan. What she did is typical of what is probably going on in many states in the aftermath of the hurricanes.

Each state faces different possible disasters. Arizona's particular challenges are the following:

- Vast wildland forest fires such as the "Rodeo-Chediski" fire of 2002, forcing about 90,000 people to leave their homes.

- An explosion at the Palo Verde Nuclear Generating Station would challenge the Phoenix area mass evacuation.

- A California earthquake could send hundred of thousands to Arizona for shelter, just as those in the Gulf Coast fled to other states.

- A dam break could trigger massive flooding, cripple communication, transportation, housing, and health services.

During the last 40 years, the federal government has declared disasters in Arizona 19 times, and 16 of those were caused by severe storms and flooding. The state has not put its head in the sand and ignored perils. Arizona has become one of four states to gain a special emergency certification by the federal government. State officials have taken part in several training events involving disaster scenarios.

Arizona was one of the first states to develop a comprehensive homeland security plan. It created a counterterrorism center that can also be a command post when a disaster hits. It has a full-time lobbyist in Washington, D.C., to deal with disaster relief, homeland security, military bases, and immigration.

Furthermore, Gov. Napolitano wants to establish a new three-digit phone number (a 211 phone line) that will let residents find temporary housing after forest fires, connect people with social services, and build an information network to track trends in bioterrorism.

San Francisco Mayor Gavin Newsom used Katrina to help citizens beef up preparations for an earthquake. More earthquakes are predicted along the San Andreas and the Hayward faults and a magnitude of 6.7 or more can produce severe devastation. Newsom warned citizens to participate in the CERT training and to keep emergency kits handy because they will probably be on their own for the first 72 hours after such a catastrophe. Meanwhile, the city is making improvements to bridges, hospitals, sirens, loudspeakers, and reservoirs to supply fire-fighting water needs.

Hurricane Katrina Prepared People for Hurricane Rita

It's a very good thing that authorities began to review their own disaster plans because less than three weeks after Hurricane Katrina, Hurricane Rita followed. Leaders from local to state level were much readier in Texas, Louisiana, and Mississippi for a new challenge. Even President Bush and FEMA came out higher in polls with their reactions to the second hurricane. So was the public readier and an unprecedented 90% of people evacuated the areas and even larger percentages were evacuated in New Orleans where some had begun to return.

However, cities like Gulfport, Mississippi, began to face a new problem when Hurricane Rita threatened the Gulf Coast after Hurricane Katrina. Were they going to be forgotten? Their homes, jobs, and businesses had been washed away. The dead and missing had not even been calculated yet. Their neighbor, New Orleans, 90 miles away, had been the focus of news and donations even though it had dodged Katrina, but it flooded because of broken levees and decades of bad decisions.

Coastal Mississippi cities had already counted over 200 dead, one-third of their homes destroyed, and many more damaged. Now Texas cities as well as more Louisiana cities and towns would be the focus of attention. Would they be remembered with donations, funding, support, social service agencies, and all the things needed to return to a normal way of life?

Local government in small but hard-hit coastal cities was still working out of makeshift buildings. Gulfport's 12 casinos on the coast were closed, leaving 17,000 employees without incomes, and those requesting financial assistance stood in long lines.

Meanwhile, a multitude of problems faced waterways. The Port of New Orleans, Louisiana, and Gulfport, Mississippi, were closed after Hurricane Katrina. The Atlanta Field Office was given oversight authority on all cargo clearance and designated the Ports of Baton Rouge and Gramercy, Louisiana, to manage transactions for New

Orleans. Atlanta designated the Port of Mobile, Alabama, to manage transactions for Pascagoula, Mississippi. Secretary Chertoff authorized the transportation of petroleum. The vessel repair entries normally submitted to New Orleans were submitted to Atlanta or to the Newark/New York area. Without river transport and waterways, the economy of the Mississippi River states is crippled.

Port cities had serious problems for many weeks after the hurricanes. People on the scene in Gulfport said that most houses were damaged, many completely leveled, and every business suffered damage. It was also hard to get around town because streets were littered with debris. Cars trying to drive down the roads suffered flat tires from the nails lying about in plywood. There was no communication because the cell towers were knocked down, so cell phones and other phones didn't work. The only people with radios were emergency-response personnel. Additionally, people could not obtain fuel so they rode with others who had gas. Only the future will tell whether the small cities along the coast are forgotten in favor of the larger cities with more dollar impact on the nation's economy.

Before Rebuilding, Improve Race Relations

Just as the traumas of wars fade, the terrorist attacks of 9/11 fade, so will the Katrina nightmare fade. That is unfortunate because improvements must be made, but it is human nature to forget unpleasant events, as we try to escape the pain of feeling and remembering terrible things. Human nature is all we have to wage war against Mother Nature. When such a disaster is fresh, astute officials will make plans to secure their own environment. Only those who are fatally in denial will pass through the Katrina-Rita aftermath by doing little or nothing.

The New Orleans flood exposed other problems which must be dealt with. How can we prevent the oppression of blacks by whites? That was the surprise that the public learned when most of the victims in New Orleans were poor blacks living in lowlands. Why did the blacks shoot at some of the law enforcement officers, and even fire and emergency rescuers who came to save them in flood-ravaged New Orleans? Have racial tensions remained so bad after all these years of working to improve them?

If you're white, put yourselves in the shoes of a black. Here's what they know about their lot in life. Three hundred or more years of slavery, torn from their roots, shackled in chains to be shipped to America, used and exploited on plantations and cotton fields in the South, later further used and exploited for their labor in the North, treated as subhuman, lack of jobs, substandard living, substandard schools and so on and so on. Frustration compounded frustration. They had separate facilities, separate schools, separate travel accommodations, separate restaurants and stores, separate drinking fountains, and so on.

Anyone old enough has seen many of these things, and most caring and thinking people have been revolted by them. It was there, though, and not enough was done by white government leaders. And New Orleans has shown that race relations are still a gigantic problem.

Blacks have had to look at life so differently than whites. The great James Weldon Johnson (1871-1938) described some of the problems faced by parents who knew that the future for a black child was more limited and difficult than for a white child.

> Awaiting each colored boy and girl are cramping limitations and buttressed obstacles, in addition to those that must be met by youth in general; and this dilemma approaches suffering in proportion to the parents' knowledge of and the child's innocence of those conditions.
>
> Some parents up to the last moment strive to spare the child the bitter knowledge. The child of less sensitive parents is likely to have this knowledge driven in upon him from infancy (56).

The lack of freedoms encountered by blacks throughout their lives can be handled constructively by people like James Weldon Johnson or vengefully. Eldridge Cleaver wrote in *Soul on Ice* that every time he had sex with a white woman, it was like a moment of freedom and revenge against white males who constrained him. These emotions have built up over the generations in the hearts of blacks. So is it any wonder that they did not see whites as rescuers? This awful condition between blacks and whites must be fixed even more than the levees must be fixed.

9

Are We Stupid to Rebuild New Orleans?

○ ○
"Luck is a mighty queer thing. All you know about it for certain is that it's bound to change."

—*Bret Harte*

There will be many opposing views about whether New Orleans should be rebuilt. The price tag for Hurricanes Katrina and Rita is already estimated at over $200 billion and may well exceed that. There is almost a complete certainty that other hurricanes will strike the vulnerable city and surrounding area in the future. There is also a certainty that serious consideration of not rebuilding the city may hurt the feelings of those who live and do business in New Orleans. Nevertheless, dialogues have already begun on "to rebuild or not to rebuild."

What does human nature tell us about whether it should be rebuilt so that those who were flooded out can return? The chances are that many whose homes were flooded will find life elsewhere better than returning to a place where they will live in another house, one that they are not attached to. People tend to cherish a particular house, church, school, neighborhood that they have seen over many years. Many of the houses in the poor neighborhoods may not have been cherished because they were inadequate and deteriorating, but nothing else could be afforded. There will be many, however, who want to return not to their own particular house but to the city they know and love, and to a job that they value highly in that city. One of the main reasons is that it is something known, and other things are unknown.

Of course, some argue that New Orleans should never have been built where it is in the first place. That may be true but it was. The same was true of Pompeii.

Two days after Katrina hit land, on August 31, 2005, House Speaker Dennis Hastert spoke with the editors of the *Daily Herald* of Arlington Heights, Illinois, a suburban Chicago newspaper. His comments (on www.slate.com) have been widely criticized but deserve some attention because they represent a widespread attitude.

Hastert said, "This is the largest disaster, natural disaster, we have ever had…"

Chris Bailey, of the *Daily Herald* asked, "Does it make sense to spend billions and billions and billions of dollars rebuilding a city?"

Hastert replied, "That's seven feet under the sea level. I don't know. It doesn't make sense to me."

Bailey asked, "Is that a question that anyone will ask or can we not even ask that?"

Hastert retorted, "I think it's a question that certainly we should ask. And, you know, it looks like a lot of that place could be bulldozed....How do you go about rebuilding this city? What precautions do you take..."

Patrick Waldron of the *Daily Herald* inquired, "...Should we rebuild this city? Would you agree that will be part of the congressional debate?"

Hastert answered, "I think it should be. Of course, the folks from New Orleans will have their own opinion on it—we are going to rebuild this city. We help replace, we help relieve disaster. That is certainly the decision the people of New Orleans are going to make. But I think federal insurance and everything goes along with it. We ought to take a second look at it. But you know we build Los Angeles and San Francisco on top of earthquake fissures and they rebuild, too. Stubbornness!"

Hastert is only one of many who question the rebuilding of New Orleans. Others say that many coastal cities that suffer from hurricanes seem to be constantly rebuilt with federal funds, why not New Orleans? Since it was learned that the French Quarter and other tourist sights have been spared by the flood, there has been a real push to rebuild New Orleans, despite the toll.

The New Orleans Metropolitan Convention and Visitors Bureau has issued statements that Katrina cost the city between 100 and 200 major conventions. Even so, they estimate that the tourism industry will continue because the French Quarter and many hotels came through the storm with little damage.

Why Should New Orleans Be Rebuilt?

While the tourist may be more concerned with the charm of the city, businesses know the importance of its port. The Port of New Orleans handles over 130 million tons of cargo a year and is one of the largest and busiest ports in the world. About 5,000 ships from nearly 60 nations dock at the Port of New Orleans every year as well as around 50,000 barges that use the waterway.

New Orleans can hardly be rebuilt without paying considerably more attention to flood control and to evacuation plans. For humans to do battle with Mother Nature is a herculean task, even greater than humans doing battle against giant creatures. Some have compared man's battle against nature to the famous novel *Moby Dick.* Herman Melville's main character, Captain Ahab, did battle with nature in the form of a whale. Ahad lost his life in that battle to "control nature." Can New Orleans do any better? If New Orleans is rebuilt, and it probably will be, let us hope that the mistakes made in the past will not be made again.

On September 12, 2005, John Schwartz, Andrew Revkin and Matthew Wald wrote an article in the *New York Times* called "In Reviving New Orleans, a Challenge of Many Tiers." They commented that before New Orleans starts rebuilding, the protections against future floods must be patched and strengthened. This will not be easy, quick, or inexpensive.

FEMA was focused on draining floodwaters, repairing the water system, getting the sewage system operating, and restoring electricity. Germs such as E. coli and others are far beyond a healthy level. It is assumed that the whole city is contaminated and unsafe for habitation. In fact, Environmental Protection Agency (EPA) spokespersons said that New Orleans would not be habitable until it has a water supply and sewage system.

The mold issue seems to have doomed most structures, especially those that have been submerged. Those returning home are finding that they have trouble seeing and breathing inside their moldy homes, businesses, and hospitals. Mold cannot be removed unless it is in the very first stages so that wide-scale demolition will be the only answer to many areas of the city and its suburbs. Of course, the same will be true in other areas besides New Orleans.

Meanwhile, how will people get in and out of New Orleans? Louisiana needs $1.3 billion in repairs to the highways according to officials of the Federal Highway Administration. Without adequate highways, railways, and other vehicle conduits, there can be no operation of schools, police, and fire stations to serve the community.

The health care system, in addition, was knocked out of commission. Katrina destroyed clinics, hospitals, medical offices, and the few hospitals still open could only offer a few beds. Kindred Hospital was used to evaluate the city's public health needs immediately after the storm. Even as it restores services, New Orleans must wonder whether it can withstand another storm after Katrina and Rita.

Some geologists speculate that it won't do any good to start building anything too soon because it is too hard to tell what to build and how high to build it. Seas are predicted to rise as global warming melts glaciers and within 10, 20, 30 or 40 years, levees and sea walls could be much less able to hold back the water.

Others will argue that because New Orleans is below sea level and sinks at a rate of one inch every three years, it shouldn't be rebuilt. New Orleans, of course, is not the only city that might be affected by glacier melt and rising sea levels. Other cities that would be affected include Galveston, Miami, New York City, Amsterdam, Rotterdam, Venice, Bombay, Shanghai, etc. (The National Oceanic and Atmospheric Administration said that September 2005 was the warmest September recorded since 1880 and that U.S. temperature for the month was 2.6 degrees above normal.)

How can such a vulnerable city be rebuilt safely? Experts seem to agree that it needs to be rebuilt to withstand a Category 5 storm. Rebuilding should certainly use the advice of those with extensive experience with flood protection systems such as the Netherlands or

England. Both Amsterdam and Rotterdam are below sea level, as well as much of Holland's countryside, and their anti-flood measures have been extremely successful. Their tall well-built embankments are reviewed and re-examined periodically to ensure their adequacy. In fact, the Dutch government announced after Katrina that they plan to spend a considerable sum (nearly $4 billion) over the next ten years to improve their flood controls.

New Orleans planners will need, say many pundits, in addition to levees, dikes, sea walls, and pumps, engineering and landscaping to prevent further erosion of the wetlands buffer that already exists. While these measures have been used in the past, they obviously require reconsideration, rather than just reenforcement.

Many would agree that rebuilding should place economically disadvantaged inhabitants onto higher and safer ground. Not only should the flooded, disreputable housing be bulldozed, but decent living quarters in communities patrolled and cared for by social services and first responders (police and fire agencies) should be available to residents.

Builders will probably include those things which have always attracted people to New Orleans. The arts, entertainments, cuisine, and history, added to the special international character of the city, may still be used to attract business and maintain tourism so vital to their economy. However, travelers as well as residents will worry about health hazards and this must be addressed in a revitalization plan for the city. Some have already suggested that these health hazards could be dealt with in creative ways, using creative architecture and energy use that would take advantage of being near a major portion of the country's oil and gas production.

One challenge will be to rebuild the city and improve the flood control system, and that will be monumental. Another challenge will be to avoid the poverty of the parts of New Orleans that leave the poor, the sick and the elderly unsafe and without help in climatic catastrophes. The primary duty of government is to provide safety for its citizens.

There will be the challenge of organizing city, county, and state emergency and evacuation procedures more effectively.

World Changes Will Affect New Orleans, Too!

There are many world changes going on that are likely to affect the rebuilding of this important port city. Data abound on the Internet and numerous books are being written about these major changes.

Since 1984, the global fish harvest has been dropping, and so has the per capita yield of grain crops. Also, ozone is being depleted; erosion and aridity is reducing nature's biological productivity; contamination of soil and water is jeopardizing the quality of food; oil, gas, and timber resources are being consumed faster than they can regenerate; and biological diversity is being lost.

At the same time, the human population and its demands on nature are growing. Between 1950 and 1990 alone, the wood harvest doubled, fish catches increased five fold, water use tripled, and oil consumption rose nearly six-fold. World population grew from 3.6 to 5.3 billion, registered cars increased from 250 to 560 million, energy consumption nearly doubled, truck transportation and waste generation has increased, and human consumption has come to exceed the global productive capacity of nature. Local human populations have frequently collapsed when resources productivity declined too steeply.

The rapid population decline on the Easter Islands around 1680, plague waves in Europe, the Irish Potato Famine in 1845, the Chinese famine during Mao's Great Leap Forward, and the chronic famines on parts of the African continent since the early 1980s are prominent examples of events where carrying capacity of a society has contributed to human population collapses.

The last forty years have been characterized by rapid human expansion and increased per capita consumption. For example, the growth rate of the car population is about double that of the human population. A global population of 10 billion is expected by 2030. In the U.S. and Canada, energy consumption is still on the rise and resource depletion has not been slowed. The latter is evident in the collapse of the North Atlantic cod fish stock affecting the Canadian East Coast, and in the forest land-use conflicts on the North American West Coast.

In summary, ecological deterioration and the parallel growth of human activity mark a sharpening conflict. Many international and local efforts have tried to help mitigate this conflict without much effect; the gap between human demands and nature's supply is widening.

Who Will Be In Charge of Building New Orleans?

Who will define and design the rebuilding of New Orleans—local, state or federal authorities or some new agency or Gulf Coast Reconstruction Czar? A *Wall Street Journal* article in September 2005 quoted Mayor Ray Nagin as saying, "I don't want anyone outside of New Orleans telling us how to plan this city!" He has proposed a commission of perhaps 16 members, eight white and eight African-American, even though the city (before Katrina) was at least 75% black. President Bush emphatically stated that the federal government will defer to the locals in rebuilding.

Meanwhile Louisiana Gov. Kathleen Blanco has stated many times that rebuilding New Orleans should not duplicate its earlier problem-filled status but should improve it.

In Washington, Congress is bracing to spend over $200 billion to repair and rebuild the Gulf Coast. They expect much say in the use of those funds to rebuild New Orleans and other areas. In fact, Congress is setting up plans for a large new federal reconstruction authority. Congressional leaders like Sen. John McCain are already proposing to axe other programs due to the unexpected huge cost of Katrina and Rita.

There may be some competition between the Gulf Coast states for government funds. Mississippi and Louisiana may both feel entitled and may offer political support for what they believe is their fair share of the funds. One New Orleans councilperson has gone on record in saying that the federal government is not including or communicating with local politicians. The Army Corps of Engineers takes the lead in designing and building major levee projects, then turns them over to the local districts for maintenance so they generally have a say in the use of federal funds for engineering.

The Orleans Levee District is an eight-seat board and six seats are appointed by the governor, who often uses the posts to reward political supporters. The board also operates the chronically money-losing New Orleans Lakefront Airport on the shore of Lake Pontchartrain, two marinas, and leases property for a casino.

Three months before Hurricane Katrina, the board dedicated a Mardi Gras Fountain with timed and colored water sprays. The fountain and plaza cost more than $2 million, partly using tax revenue that the board collects for levee maintenance.

Mayor Nagin has said that Louisiana received money from Harrah's Casino to build state edifices in Baton Rouge.

An Epic Evacuation Made Way for Hurricane Rita

It would be interesting to know where things would have landed if the hurricane season had ended with Katrina. But it didn't. Within three weeks, Hurricane Rita developed and became a category five, then a category four, but by the time it hit the Gulf Coast had been downgraded to a category three.

On September 23, 2005, less than a month since Hurricane Katrina, Texas Gov. Rick Perry prepared for Hurricane Rita to hit the coast of Texas. Evacuations were ordered by coastal cities, and airlifts, buses, and cars were transporting passengers to inland cities. The news reported that long traffic lines to evacuate were causing several problems. Cars ran out of gas while idling in heavy traffic on highways. One bus transporting elderly and sick patients, many of whom had oxygen tanks with them, exploded and many were killed.

However, Gov. Perry told reporters that because so many took evacuation orders seriously, they would get through the disaster with the help of thousands of rescue and relief workers standing ready. It was probably thanks to the photo coverage by the media that the public chose to evacuate rather than risk landing in predicaments they saw on television.

Perry cited the lessons learned from Katrina when he issued a disaster declaration in anticipation of landfall in Texas. He asked President Bush to approve the disaster declaration and 100 percent reimbursement for Texas communities responding. He also recalled Texas National Guard, Texas Task Force 1, and all other emergency personnel and equipment from Louisiana to prepare for Hurricane Rita in Texas. He ordered the activation of 5,000 Texas Army National Guard personnel to support preparation efforts, including 11 helicopters. Some 500 Texas State Guard members were activated to assist the American Red Cross with shelter management.

Texas prepared immediate care strike teams consisting of the American Red Cross, the Salvation Army, and the Texas Army National Guard to move rapidly into the area where the storm was to strike. Governor Perry warned coastal residents to get out two and one-half days before the hurricane was to hit land.

Mandatory evacuations were ordered by the mayors of Galveston and Port Arthur, Texas. "We've always asked people to leave earlier, but because of Katrina, they are now listening to us and they're leaving as we say," said Galveston Mayor Lyda Ann Thomas.

Evacuees who had been taken to Texas after Hurricane Katrina hit the Gulf Coast were being moved again, this time to Arkansas.

Hurricane Rita also made a glancing blow to New Orleans with winds and rains that broke through temporary levees erected only days earlier. Louisiana Governor Blanco declared a state of emergency four days before landfall because engineers had predicted that even a foot of water could swamp the city's levees.

New Orleans asked for 200 buses to help in a possible evacuation. Residents who had just returned home were told to be ready to evacuate again. The city was closed two days before Hurricane Rita was to hit land.

Just before Rita hit, Louisiana's senators introduced legislation that would provide an estimated $250 billion in federal money to rebuild flood-ravaged New Orleans and repair hurricane damage elsewhere across the state. Congress already had provided $62 billion for recovery of the Gulf Coast from the hurricane and previous estimates had suggested the final total could top $200 billion. This didn't include reconstruction funding for Mississippi or Alabama, which also were hit hard by Katrina. Nor did it include the additional costs generated by the second hurricane—Rita.

That was only one state of the several that were hit within a month by hurricanes. Who is killing New Orleans and America? Are the perpetrators Mother Nature with predictable disasters or human nature? All we have to control Mother Nature is our human traits. Should our leaders make better decisions by using more foresight? Should they believe in and prepare for known facts and future disasters? Are we stupid to rebuild New Orleans without taking care of the causes of its ruin first?

10

What Can We Learn from the Past?

"Those who cannot remember the past are condemned to repeat it."

—*George Santayana*

Mankind keeps changing not only the way we live but where we live and work. First, early man was a hunter and gatherer. Some ten thousand years ago, he began to settle down to grow crops and domesticate animals. More permanent settlements, villages, businesses, and trade routes sprang up along major waterways. Those waterways included the Nile, Mississippi, Indus, Tigris-Euphrates, Yellow River, Yangtze, Amazon, and other major rivers of the world.

Boats, sailing ships, and all sorts of river transport led to world trade systems that gradually connected all parts of the world. As other methods of transportation arose, horse-drawn vehicles, covered wagons, railroad trains, automobiles, airplanes, and now communication brought the world ever closer together.

In the past, people used to grow up and live their entire lives in one small community. With our mobile society, that time has passed. Nowadays we move about according to the jobs we take and the preferences we have for our surroundings and our relationships.

We generally tend to live near people we are like or near those we'd like to become more like. Usually people with the same values live fairly close to each other. Businesses and stores build up around our homes because they need customers and choose where they have customers. They offer the services and products that are needed by the people near them. That used to be truer than it is now because we can order products and use services of businesses located much farther away from our residence, thanks to mail order, telephones, internet, trucks, railroads, planes, etc.

All of these features allow us many more choices about where to live and where to do business. My point is that nobody in the United States is really stuck living and working in or near the flooded areas of New Orleans, the San Andreas earthquake faults, the California mudslide areas, the Mohave Desert or the Mount St. Helen's volcano. People choose to live in places that are at sea level and sometimes, even below sea level. They live close to large bodies of water and/or oceans.

They live on hillsides that are prone to sliding. They aren't forced to live there—they have chosen to do so.

Some might say that poor people or the poor blacks of New Orleans can't afford to move so don't have a choice. That would be wrong, of course. The Dust Bowl disasters in the United States in the 19th and early 20th century caused thousands of poor farmers to flee from the Plains States to California and other more fertile areas. Migrants' decisions to flee were only one of a variety of survival strategies pursued by poor families along with other strategies such as selling possessions, eating different foods as they changed location, and taking new kinds of work.

In 1923, the U. S. Labor Department estimated that some 500,000 blacks a year were going North to get away from the active Ku Klux Klan and oppressive Southerners. The Klan was at its height and had helped elect some 16 senators. Blacks sought jobs in the growing Detroit automotive industry and many factory jobs in the Northeast. That massive migration was captured in the 1923 song *Bye Bye Blackbird* and its poignant phrases described earlier. The poorest of the poor blacks from World War I up to and through World War II were able to flee oppressive anti-black attitudes in the South. They did not even have the estimated average annual income of $32,000 of the New Orleans 9th ward black residents. Individual choice and free will are at work for almost everyone except those disabled who must be cared for by others in places not of their own choosing.

The concept of individual responsibility has been growing in the field of emergency medicine. A recent development in emergency services is to charge a fee for some rescues. In the past, those who violated signs or orders or did stupid things that required rescue workers to put their lives on the line got free rescues.

Increasingly fire services and other search and rescue teams believe that their danger should be compensated. Actually, the theory behind rescue compensation is that it will provide a deterrent to those that are considering a "stupid" move. It also can provide funding to support the rescue teams' equipment and training. Did you get yourself into trouble rock climbing and need a helicopter rescue? Ever drive into deep water on a road despite signs advising against it and find yourself unable to get out? Ever go hiking on the Grand Canyon trails without taking the amount of water recommended by rangers on trail signs? Don't worry, you'll still be rescued, but you had better be ready to pay for it these days.

The chances are, however, that most of the people who ignored evacuation orders during Hurricanes Katrina and Rita will not be charged. Yet officials had to maintain hundreds of boats and rescue vehicles to save those who stayed behind and tried to tough it out.

Individual responsibility is akin to societal responsibility. Responsibility is a theme of scientist Jared Diamond who has an important addition to the equation of who kills a city. He believes that politicians and business leaders make decisions that save or wreck civilizations and cities. Archaeological discoveries have revealed that natural resources were thoughtlessly exhausted and not replaced, leading to the collapse the Maya in the Yucatán, the Anasazi in the American Southwest, the Cahokia mound builders outside St. Louis, the Greenland Norse, the statue builders of Easter Island, etc.

Cities Can Choose to Fail or Succeed

Jared Diamond proposed in his new book *Collapse: How Societies Choose to Fail or Succeed* that societies choose to kill themselves by their own governing decisions. So might it be for cities. He blames aggressive and profit-centered companies, agricultural and resource mismanagement, and bad governance by foolish leaders and the greedy ruling elite.

Diamond shows how some societies manage to reverse what might otherwise be a crucial mistake when beset by a range of problems and mounting failures. He pleads for us and our city leaders to oppose decisions that will lead ultimately to the suicide of our cities. He noted that governments often have a short-term or 90-day focus, and mainly respond to crises. Sometimes crowd psychology pushes governmental authorities toward quick reactions as they try to maintain public approval. This is likely to reduce critical thinking and suppress painful realities just when they might be needed.

If leaders seek "yes" men and ask colleagues to minimize disagreements, as Bush and other leaders are accused of doing, they are likely to run into problems that nobody discussed in meetings and briefings. But if they ask colleagues to think critically and ask questions, they are likely to make better decisions and be more prepared for worst case scenarios.

The new problems being faced in America and the world require much forethought. When we have a combination of consuming resources, degrading our environment, producing endless waste, that is a tough enough set of challenges. Then add to that the unexpected climate changes seen in Hurricanes Katrina and Rita and decision makers are incredibly challenged. Will they plan for the short-term, for their own special interest groups and pet projects and profits? Will they plan for the long-term, for the survival of their children, by maintaining and replacing and caring for their surroundings? Will they build and restructure New Orleans to protect its citizenry? Will leaders take expected climate and natural resource depletions into account as they plan for our future? Will leaders think technology will solve our problems and proceed to spend our resources recklessly ignoring conservation practices?

We think we are looking ahead as we plan our cities. Didn't the ancient political leaders look ahead and recognize their problems? Their attention was evidently focused on the short-term concerns of gaining wealth, winning wars, erecting giant structures, outdoing each other, and producing enough food to meet their immediate needs. Sound familiar?

What else captures our attention and distracts us from long-term thinking? We have more crowding, commuting delays, water deficits in arid states, forest fires from forest mismanagement, air quality problems, decreased fisheries, higher gas and energy prices, expensive toxic cleanups, unwanted pests, indestructible germs, terrorism costs, illegal aliens, and increasing budgetary deficits. Additionally, we are interconnected to the rest of the world and are involved in a complex system of political and economic alliances, mutual exchanges and protections, and trade agreements. We will never ever be able to separate ourselves from the rest of the world.

How can we recognize problems when things and conditions fluctuate widely up and down over time—like climatic changes and financial indicators? If and when we recognize a problem, do we solve it for the short (political) term or for the long-term, like during our children's lifetime? Do we make or save money with our decisions or do we bite the bullet and spend what's needed to prepare for the future?

Diamond pointed out that not all societies make short-sighted and fatal decisions. There are parts of the world where societies have not collapsed, such as Java, Tonga, and (until World War II) Japan. He noted that Germany and Japan are successfully managing and even increasing their forests, as is the Dominican Republic. Alaska's salmon fishery and the Australian lobster fishery are being managed relatively well. If a city or society operates with a minimum of differences between its leaders and its citizens, it is more likely to prepare well for the future and survive. When the leaders see things one way and the citizens another, it can make for the biggest problems.

Politicians and leaders have special access to information, must make decisions, and can redistribute surplus goods. This centralized leadership and decision making enables them to reward themselves and their supporters, sometimes at the expense of the general welfare of their societies. Throughout history, these were the common characteristics of authoritarian governments. However, even the more democratic and open societies can be tainted by these characteristics or coping traits.

Diamond cites examples of the executives of Enron, Tyco, and Adelphi who selfishly took from company coffers. A society that minimizes such differences is the Low Countries, whose citizens have perhaps the world's highest level of environmental awareness and of membership in environmental organizations. Much of the Netherlands is below sea level, and was reclaimed from the sea using dikes and pumping out the water. Everyone would go down together if the dikes and pumps fail.

People find it hard to realize that many places on the earth are very dangerous and that they should not risk their lives by living in some places. People place themselves in the path of danger all the time. Why? They seem to live with the idea that nothing can happen to them. They operate with that very bad coping technique called denial of reality.

Each person has the choice of choosing where to live and upon what to build their homes and lives. Each person has the choice of listening to the disasters and sounds of Mother Nature saying "respect my movement" and "stay out of my way." Should we stupidly pick up the pieces and continue, disregarding the warnings that this disaster can happen again? Natural disasters will continue to happen. It is only a matter of time as to when they will strike again.

Disasters are natural events, not the end of the world. How we deal with them and make choices will determine how much suffering we go through in the future. We cannot control Mother Nature but we can control what we do about it because we have free will.

These questions need to be addressed more than blaming others or God or any government once something happens.

Are We Biting Off More Than We Can Chew?

Population displacement due to the environment is not new. Historically and prehistorically, people have moved because of natural disasters, wars, or having too many people for too few resources. Cities have been abandoned (Babylon is an example) because they were badly located. Environmental disasters are displacing more and more people every year because we live in such densely packed communities that they require little disturbance to be disrupted. People and governments have changed their physical environment so much that it has become more vulnerable to disruption. Denser populations consume more trees, natural resources, water, generate more waste products, and contribute to global warming.

Some have said that man was not meant to live in large cities. Large, dense populations outgrow the resources in their location. The resources we need—water, agriculture, mineral resources, transportation, and all their accompaniments make us less able to survive even some minor disruptions or disasters that occur (loss of electricity, loss of communication systems, loss of fuel for automobiles, loss of food, loss of water, loss of sewage and refuse services, loss of professional health personnel, etc.)

As if the density, the need to protect communities from the environment, and the overuse of resources aren't enough, our large cities face other problems:

- the flood of immigrants over national borders,

- indecision about whether to increase, preserve, or use up resources such as oil, gas, trees, etc.

- indecision about whether to change eating habits since some food sources are being exhausted (many species of food fish are disappearing, cow diseases are minimizing beef, fowl are being killed to prevent avian flu, etc.)

- salination of soil and contamination of freshwater marine food supplies by floods of ocean saltwater

Again, these questions need to be asked at every level of our society when things are calm rather than simply blaming others after a disaster happens.

Actually, immediately after a disaster such as Hurricane Katrina or Rita, or any earthquake or natural disaster, an important healing factor is for city residents, local government, businesses, and organizations to resume normal activities. But, therein lies a problem. In the rush to re-build, local officials and citizens must understand why the damage occurred and then consider how to reduce a city's vulnerability to the next disaster.

It can be argued that the state has a duty to protect its citizens from disasters. If the state is negligent or indifferent to meeting its obliga-tions to protect its citizens' basic needs, it could be considered a breach of the basic contract between citizens and their government for which they pay taxes. However, the local, state, and federal government's responsibilities with respect to those in humanitarian need are not well-defined. People displaced by environmental changes and disaster victims can hope for help from their government, groups, wealthy donors, and insurance, but there are no defined compensation pack-ages. Should there be? That is another issue requiring discussion rather than blame.

How Can Healing Occur?

Resuming normal life is important for helping people to recover fol-lowing a disaster. Surely compensation packages are one way to help. Another way is to have temporary programs that allow citizens to assist in the reconstruction of their communities. Work projects can involve cleanup efforts, housing reconstruction, and rebuilding of the commu-nity destroyed or damaged in the disaster. A community that becomes involved in its own reconstruction programs can reap beneficial psy-chological effects. Men, women, and children should be involved in rebuilding a community so that it represents the interests of those who will live there.

During reconstruction, the community can become involved in reducing its susceptibility to hazards by being informed of how build-ings and environmental factors interact. Communities can create refor-estation, dams, stabilize hillsides, improve drainage systems and waterways, remove structures in flood zones, and use engineering that will create a more sustainable community. Communities can reduce their vulnerability to disasters by incorporating self-monitoring methods to be sure that safe building practices are being followed.

Obviously, cities must relocate people away from land that is vulnerable to disasters. In relocation projects, dwellings and businesses should be roughly comparable to those used previously, accessible to jobs and services that fit the needs of dwellers, but less vulnerable to disaster risk. They also need to permit people to choose where to live within the relocation project.

Drinkable usable water must be provided by installing adequate sewage systems before construction begins. Water in today's world must be transported from ever greater distances. Years ago, you couldn't live anywhere without water being close by. Now people choose where to live and they usually don't ask if there is a 100-year water supply available. The desert area around Phoenix, Arizona, is a good example of a rapidly growing metropolis in an area with scant water available.

Only in the last two or three generations have people had washing machines, swimming pools, lawn sprinklers, and daily baths and showers. These water consumption habits are more difficult to sustain in some areas than others depending on the aridity.

The Long Battle between Mother Nature and Human Nature

Many do not understand that our earth is a living breathing organism, just as we are. The Earth has been constantly changing since its creation in space and time. It huffs and puffs and spits with geysers and volcanoes, it convulses and sweats and rains with floods and tsunamis, it shrinks and expands with earthquakes, and it has chills and fevers as ice caps freeze and melt. It gets sunburns in the form of sun spots and solar emissions. "The earth moves under our feet," as the song says.

Our earth has been changing since the beginning of time. Not so long ago, cities like Chaco Canyon in the New Mexico desert had water, but then it dried up and people moved away. Not that long ago, pioneers were afraid to ride the rough waters through the Grand Canyon but they are easily navigable now. Water tables rise and fall as does the melting and freezing of water on mountains.

Early man knew that weather conditions changed and they thought they could influence those changes. They prayed, chanted, danced, sacrificed, and used all manner of methods to influence the living earth to remain safe and to supply his needs. Modern man is farther from nature, at least in the United States, and spends most of his day indoors thinking about almost everything but his safety and his earthly supplies. So disasters may come as more of a shock.

We can no longer think of disasters as being "over there" or somewhere that has no connection with us. National and world news brings world events closer to each person who watches a television, listens to the radio, reads a newspaper, or turns on the Internet. The photos of any event are quickly available and can be quite disturbing. The humans of the earth are now a worldwide family.

We are fortunate in the United States to have such sophisticated police and fire systems. Virtually every community in the country has been evaluated by their emergency management systems including local government offices for possible risks. Scenarios and practices prepare first responders regularly for all sorts of disasters.

As communities rebuild after hurricanes, risk assessments or hazard mapping may be valuable to prevent future disasters. Such assessments may have already been prepared by your city government, fire departments, and police services and may include information about:

- residents (ages, vulnerable groups, health level, ethnicities, traditions, social values, attitudes to hazards, trades, and disabilities),
- communication networks,
- mobility and transportation networks,

- waterways, landform distributions and geology (climate, disaster history, flora and fauna, equipment and vehicles for urban and rural hazards of the area),

- agriculture and livestock,

- hazard awareness,

- hospitals, shelters, helicopter and airplane landing locations,

- essential services (utilities, backup generators, emergency systems)

- vulnerable structures (old buildings or homes, dams, levees, bridges, oil refineries, skyscrapers, depositories for dangerous materials)

Water supply and sanitation systems can be designed or redesigned to reduce vulnerability. Good systems and backups can not only enhance the fire personnel to suppress fires but can reduce the risk of epidemic diseases such as cholera, real risks during the floods in New Orleans.

An additional degree of safety can be provided if community members discuss vulnerabilities and hazards with those responsible for local water or sanitation. Just think how valuable it would have been if New Orleans flood victims could have had water to drink.

Traditional means of water purification and traditional rules for segregating water uses (stock watering, bathing, drinking, golf course watering, swimming pool sanitation, etc.) can be discussed with residents when rebuilding a community. Then residents can assist local authorities to plan, maintain, and reduce vulnerability for their community. In fact, these discussions can often lead to low-cost improvements and the basis for local level emergency planning. Will this happen in rebuilding New Orleans and other flooded communities in Louisiana, Mississippi, Alabama, and Texas?

Open meetings and public education are so needed on treating household water (filtration, chlorination, boiling, storage, etc.), oral hydration therapy, breastfeeding, food hygiene, hand-washing, the use of latrines, waste disposal, drainage, and vermin control to prepare a community before any emergencies occur. Refuse, waste management and storm drainage information can help communities plan what to do and where to do it when money and labor are in short supply.

Participatory planning must involve people from the start of the planning process. Does Ray Nagin anticipate this kind of input in rebuilding New Orleans? It is not enough simply to ask people for their opinion of a plan that has already been drawn up. Leaders must listen to people in the communities where disasters have happened.

This is not always easy because planners and people in the community often have different priorities and perceptions. For example, an environmental planner may think that improving drainage is the most important thing, while the residents may view the prevention of crime or malnutrition as more urgent. Drainage, or at least the health problems and possible flood risk produced by poor drainage, is probably one of the priorities of residents as well, but not the first priority.

Are We Condemned to Repeat the Past?

One of the keys to improving emergency preparedness is the ability to learn from previous disasters and to incorporate that learning into practice. However, many of the lessons are *not* learned because staff turnover disrupts the memory of a disaster. And there is a strong desire to re-establish the "normal" state of affairs as soon as possible after an emergency. People are already resigning from public positions in New Orleans (see Police). Who will remember what happened in 2005 when plans to maintain things are reviewed in 2008? How can we prevent future disasters when we rush to re-establish "normality"?

Learning from the past could be helped by establishing a periodic review of emergencies and disasters with government officials and residents. However, it must have support at the highest level. Otherwise, the memory of disasters can be lost in the "noise" of competing bodies.

Hopefully, records kept of Hurricanes Katrina and Rita will include a detailed account of how re-engineering or rebuilding levees came to be made, what the expected benefits were, and what the present consequences are.

The Answer is Better Personnel Selection!

The biggest thing that we can learn from the past is that human nature is our only weapon to battle with Mother Nature. We must therefore redouble our efforts to select only the best people to be our leaders, and they must select only the best people to advise them and think with them to protect us.

We must rid ourselves and our leaders of the notion that they should reward people for their political donations and support with positions for which they are not well qualified. At the highest levels of government and publicly-funded agencies, we must have people with a proven track record in the work we expect them to administer. The selection of people for the highest positions may not include using an assessment center, for only people at lower levels would design the assessment center. Selections and appointments must be based on proven past experience and character by collection of references from the widest possible range of people and previous positions of power.

Special attention must be paid to the character, the habits, the morals, and the defense mechanisms of people who are to occupy the highest positions at the federal, state, county, and city level, as well as those who run agencies as important as FEMA, EPA, FBI, CIA, Homeland Security, etc.

Good leaders will surround themselves not with those who reassure them that all is well but who anticipate worst scenarios as well as best scenarios for discussion, those who help and guide planning for the good of the people, all the people, in this country. Good leaders and good advisors will bring critical information to the table so that good decisions can be made, no matter how unpopular those decisions are.

Those who fear rejection, those who are worried about their self-esteem, those who are afraid to be questioned, and those who become paranoid if they are criticized are too weak to lead. They will succumb to the flattery of the press and pressure groups.

Those who are ambitious to lead because they are tempted by power (that great "aphrodisiac" according to Henry Kissinger) may react with anger to those who question them and their use of power. They may fear the loss of power and therefore of prestige so much that they seek only "yes" men.

Those who seek the trappings and wealth of high office will tend to over-reward those who offer them more campaign contributions and support. They will be too ready to enjoy the things that are offered to those in high positions—illegal gains, extramarital sex, more luxurious lifestyle and possessions, assurances for future benefits in return for favors, etc.

Let us vote for and select people who are altruistic and whose experience proves that they are not selfish but rather selflessly help others. Let us seek those who anticipate and plan for problems, and who are willing to practice for emergencies by training exercises that involve leaders at the highest levels, instead of just subordinates. An annual practice for disasters is not enough for our president, cabinet members, governors, mayors, and executives of organizations like FEMA, FBI, EPA, Homeland Security, etc. Not only do the disasters caused by terrorists loom in the future, but natural disasters and sudden disruptions of certain resources impact people more heavily in this day and time.

In addition, let people at the highest level have the sense of balance supplied by a bit of humor, which underlies the ability to laugh at oneself. Without that ability, we take criticisms too seriously and fear making mistakes. We must be flexible and able to change courses quickly when it becomes obvious that the course being taken is not effective enough.

Let us avoid having leaders who deny facts, deny responsibility, blame others, think only of the short term, focus only on praise and prestige, and lie to cover up their bad judgment and mistakes. Let our leaders welcome new information, even if it suggests major changes and unpopular positions. Let them spend our money for the right things that foresight suggests are needed rather than the wrong things like short-term fixes in response to pressures.

The real tragedy of the Hurricane Katrina and Rita devastation was caused by the exposure of the human nature of those we elected and hired to look out for our long-term welfare. The only answers are to select and hire people with different and more effective habits, character, and coping methods that are effective in handling the stresses, temptations, pressures, and problems that leaders must handle in crises. If they aren't as intelligent and knowledgeable, let them surround themselves with those who are. Some of our best presidents have had the least formal education, like Harry Truman.

We must select leaders who will address racial inequalities, health care inequalities, prepare and protect us from attacks by nature or enemies, and lead the way to helping us change our lifestyle when it threatens to cause us worse problems in the future.

The history of New Orleans is not over and is being made each day. Let us hope that the next faltering steps are made with the greatest forethought by leaders who may not choose the most popular actions but choose the best actions in the end. They may content themselves to remember the words of Abraham Lincoln.

"I do the very best I know how, the very best I can; and mean to keep doing so until the end. If the end brings me out all right, what is said against me won't amount to anything. If the end brings me out wrong, ten angels swearing I was right would make no difference."—Abraham Lincoln

AFTERWORD
By Johannes Spreen

Hurricane Katrina ravaged the Gulf Coast with unbelievable devastation. Parts of Mississippi (Biloxi, Gulfport and other towns) are in ruin and New Orleans is in chaos.

We all heard the victims' plaintive cries, "What will we do now?" "Where will we go?"

KATRINA EXPOSED FAULTS

Hurricane Katrina, quickly followed by Hurricane Rita, exposed faults in the levee system, faults in communications, faults among political leaders, faults in police work with some police, and faults with people.

Yes, there was plenty of blame to go around at the local, state, and federal levels of government as well as government agencies such as FEMA. They were blamed for improper attention—after numerous warnings—to the costliest natural disaster in U.S. history. Over 90,000 square miles of territory in Alabama, Mississippi, Louisiana, and even Texas were declared disaster areas.

As political leaders, media, police and people discuss and deal with the devastation and search for answers, between the finger pointing, the ugly specter of racism again permeates the land.

The good and the bad in people emerge in fights, gunfire, anarchy—atavistic and feral tendencies erupt. The entire country has been on edge for weeks now—and then racism? Déjà vu all over again!

RACISM?

While those in the business of fomenting racism, usually for their own ends, continue, let's look at what really has happened. Peoples' hearts all over America went out to those victims, mostly poor and mostly black. There were tremendous contributions of money, time, effort, and support by thousands of Americans, white and black. No color line seen here!

And after Katrina came Rita—another devastating hurricane. Ironically, after the great help offered by people of Texas to the people of New Orleans, those fine Texas citizens also needed help.

After Rita, many houses in New Orleans were under water again. Mold grew all over the city—in the houses of the poor, the middle class, and the wealthy. It appears that some houses that were earlier thought to be spared will eventually have to be torn down. Sad—sad—sad!

SOME PROBLEMS—SOME ANSWERS

At the writing of this book, I have reached the age of 86, with 44 years devoted to police work—law enforcement, and many years as a professor of criminal justice.

I have coped with police problems, political problems, people and pressure group problems, and relating to and coping with the media. Katrina and Rita caused problems to abound—but the two major problems I see are the failure of government leaders to plan, lead, and direct and the failure of all of us to eliminate racism.

FAILURE OF GOVERNMENT LEADERS: BUREAUCRACY

Many could benefit from a better knowledge and awareness of administrative management principles—which I taught in college. Luther Gulick espoused the following principles of administration: Planning—Organizing—Staffing—Directing—Coordinating—Reporting and Budgeting. There were failures in all these areas in the aftermath of the hurricanes.

When I taught police administration, I highlighted the top three principles by means of a triangle to plant the concept in the minds of my students. At the three corners were the words Planning, Organizing, Directing which I nicknamed POD. I taught that planning should include budgeting—you need money to do anything. Organizing includes staffing with personnel, then training and equipping them. Directing includes commanding—someone in charge, and controlling—measures or devices as to how people are performing. If things don't quite work out, turn the corner and re-plan. Of course, each area of POD involves external relations (public relations, reporting), internal relations (good communication informs and boosts morale). Not only police but city, state and national leaders should heed these long-appreciated principles.

FAILURE TO ELIMINATE RACISM

Racism was there when I started in policing in '40s. It was certainly there when I was Police Commissioner in Detroit in the '60s. It was there when I was sheriff of Oakland County, Michigan, in the '70s, and I'm sorry to say it's still with us.

But, the help, understanding and generosity of millions of Americans of all colors, all nationalities, and all creeds have burst forth to aid and support survivors of Katrina and Rita—our fellow members of our human family. That *was* and *is* Love! Let's have more of it: love is the real answer, isn't it?

- If it's caring about your neighbor so you report an assault you witness upon him or his home, that's love.

- If it's caring about your community so that you don't want to see it suffer, that's love.

- If you care about your fellow citizens no matter what their hue, that's love.

- If you care enough to willingly serve your country and your community, that's love.

- If you are concerned about the conditions that can tempt man to harm his neighbor, and you want to see them alleviated, that's love.

- If you feel that there are things wrong, injustices, evils in this world, and you earnestly wish to do something about them, that's love.

- If you put your personal desires and politics second to your concern for your community, that's love.

- If you can take a negative and help turn it into a positive, that's love.

- If you use consideration, care, courtesy and compassion in your dealings with all you meet, that's love.

- If you live according to the Golden Rule, the Ten Commandments, or your moral, ethical, or religious beliefs, that's love.

- If you consider the feelings of the other person as an individual who is with you on this small spinning spec of dust called earth, that's love.

- If you have faith in people and in your police, that's love.

- If you have hope that we can all live together in a better world, that's love.

- If you offer charity to all your fellow men, that's love.

- If you believe there may be a spot in heaven for all, regardless of their race, color, or creed, that's not only love but heaven on earth.

NEW ORLEANS—A ROLLING CITY?

There will be other storms and hurricanes. Katrina and Rita were monster storms and more will come. In 2004 Hurricane Otto hit as late as November 30th. Will there have to be more panic, more evacuations, more loss of life? Probably. Florida, Alabama, Mississippi, Louisiana, and Texas are apparently most at risk.

New Orleans wants to re-build? It is my feeling that attached homes should be built according to strong codes to withstand category 5 hurricanes, if such a thing is possible. The levees, of course, or protective structures should withstand category 5 hurricanes. But building should not resume until the ground area is deemed safe.

The 9th ward of New Orleans, impacted so terribly, should move very cautiously. After all, who would want to invest in rebuilding that area. If it is to be rebuilt, I recommend that no more permanent structures be erected there, but I'm no engineer. It just seems to me that mobile homes and mobile offices might allow people to inhabit the area, if it must be inhabited, so that in case of hurricane, they can roll away to higher safer ground.

Whatever is done in New Orleans must take into consideration long-term thoughtful planning in two main areas: leadership to increase flood control and decrease racism.

AFTERWORD
By Bob Cheney

History is affected by geology. Each day the sea creeps up somewhere on land and cities disappear under the water. Rivers swell up and flood or dry up or change their course. Lush valleys become deserts. We know all this but we also know that the climate does not control us as much as it used to.

Mother Nature's challenges have made us smarter. We continue to develop new responses to the demands of environment and survival. We have learned to irrigate and air-condition deserts and build dams and reservoirs to control waters. But a tornado or flood or iceberg can teach us what we fail to remember, we aren't in control of Mother Nature.

We have learned some things about Mother Nature that probably apply to Human Nature as well. Competition, selection, breeding, and advantages are very important to evolution and to human beings. We might like to create societies according to our ideals but we can't, so we operate with just basic human nature. The concentration of wealth and power is natural and inevitable, and is kept from becoming too inequitable by laws, revolutions, and the power of public opinion—which is influenced by the media.

"Democracy is the worst form of government in the world, except for all the others," according to Sir Winston Churchill. It is certainly the most difficult because it requires so much intelligence. Democracy has probably done less harm and more good than other governments. But American democracy has its failings as we can see in these disasters. Democracy was supposed to break down the walls of privilege and class, and raise up ability and rank in each generation.

It is, of course, true that all men and women can't be equal but their access to education and opportunity must be made as equal as possible. We find after Katrina and Rita that our democracy is divided between the fortunate and the unfortunate. In such a state as we now find ourselves, a failure of our leadership may leave us weakened and corroded with internal strife. Have the comforts and conveniences of those who live the good life weakened our moral fiber and blinded us to the unfortunate? If we are fortunate, we must gather up as much as we can of our good fortune and share it with our fellow man.

RESOURCES

Allen, Mike. 2005. Living too much in the bubble? *Time*, Sept. 19.

Ambrose, Stephen. 1996. *Undaunted Courage*. New York: Simon and Schuster

Barry, John. 1997. *Rising Tide: The Great Mississippi Flood of 1927 and How It Changed America*. New York: Touchstone.

Blanchard, Edward and Edward Hickling. 2004. *After the Crash: Psychological Assessment and Treatment of Survivors of Motor Vehicle Accidents*. New York: American Psychological Association.

Cardini, Franco. 1989. *Europe 1492: Portrait of a Continent Five Hundred Years Ago*. New York: Facts on File.

Carey, John, Ed. 1988. *Eye-Witness to History*. Cambridge: Harvard University Press.

Carlson, Peter. 2005. After the deluge, New Orleans's Mayor Nagin stands his ground. *Washington Post,* Sept. 2.

Cleaver, Edlridge. 1999. *Soul on Ice*. New York: Delta.

Diamond, Jared. 2005. *Collapse: How Societies Choose to Fail or Succeed*. New York: Viking.

Douglas, John et al. 1999. *The Anatomy of a Motive*. New York: Scribner.

Douglas, John 2003. *Anyone You Want Me to Be*. New York: Scribner.

Fagin, Brian, ed. 1996. *Eyewitness to Discovery*. Oxford: Oxford University Press.

Faulkner, William. 1995. *The Wild Palms (Old Man)* (1939). New York: Vintage.

Friedman, Thomas L. 2005. "Katrina reveals Bush team's flaws." *New York Times,* reprinted in the *Arizona Republic*, Sept. 9.

Freud, Sigmund. 1966. *The Ego and Mechanisms of Defense.* New York: International University Press.

Gibbs, Nancy. 2005. New Orleans lives by the water and fights it. *Time,* Sept. 12.

Hastert, Dennis interview on Aug. 31 with editorial board of the *Daily Herald* of Arlington Heights, Ill., reported on www.slate.com.

Holloway, Diane with Bob Cheney. 2002. *Analyzing Leaders, Presidents and Terrorists.* San Jose: Writers Club Press.

Holloway, Diane and Bob Cheney. 2001. *American History in Song: Lyrics from 1900 to 1945.* Lincoln: Author's Choice Press.

Holmes, Margaret M. 2000. *A Terrible Thing Happened: A Story for Children Who Have Witnessed Violence or Trauma.* Franklin, TN: Dalmation Press.

Holy Bible: American Standard Version. Thomas Nelson & Sons, 1901.

Jacoby, Tamar. 1998. *Someone Else's House: America's Unfinished Struggle for Integration.* New York: Basic Books.

Johnson, James Weldon. 2000. *Along This Way.* New York: Da Capo Press.

Klein, Joe. 2005. "Listen to What Katrina Is Saying." *Time,* Sept. 12.

Kluger, Jeffrey with Cathy Booth Thomas. 2005. "The fragile gulf." *Time,* Sept. 12.

Kohlberg, Lawrence. 1984. *The Psychology of Moral Development.* New York: Harper Collins.

Kunstler, James Howard. 2005. *The Long Emergency.* New York: Grove/Atlantic Monthly Press.

Lacayo, Richard. 2005. Rebuilding ad dream. *Time,* Sept. 12.

Lewis, John. 2005. This is a national disgrace. *Newsweek,* Sept. 12.

Long, A. L. 1983. *Memoirs of Roert E. Lee.* Secaucus: The Blue and Grey Press.

McCullough, David. 1992. *Truman.* New York: Simon and Schuster.

McPhee, John. 1989. *The Control of Nature.* New York: Farrar; Straus & Giroux.

Meltzer, Milton. 1960. *Mark Twain Himself.* New York: Bonanza Books.

Pyszczynski, Thomas et al. 2002. *In the Wake of 9/11: The Psychology of Terror.* New York: American Psychological Association.

Reeves, Thomas. 1992. *A Question of Character.* Rocklin, CA: Prima Publishing.

Reynolds, W. Kirk. 1984. *Atlas of American History.* New York: Charles Scribner's Sons.

Rieff, Philip. 1963. *Therapy and Technique: Sigmund Freud.* New York: Collier Books.

Ripley, Amanda. 2005. An American tragedy: How did this happen? *Time,* Sept. 12.

Samuelson, Robert J. 2005. Hitting the economy. *Newsweek,* Sept. 12.

Samuelson, Robert J. 2005. "Why cheap gas is a bad habit." *Newsweek,* Sept. 19.

Schnurr, Paula and Bonnie Green. 2003. *Trauma and Health: Physical Health Consequences of Exposure to Extreme Stress.* New York: American Psychological Association.

Scutari, Chip. 2005. State preparing for the worst. *Arizona Republic,* Sept. 14.

Spreen, Johannes with Holloway, Diane. 2004. *American Law Enforcement Does Not Serve and Protect.* Lincoln: Iuniverse.

Spreen, Johannes and Holloway, Diane. 2005. *Who Killed Detroit: Other Cities Beware!* Lincoln: Iuniverse.

Suhayda, Joe. 2000. Life on the Mississippi. *Time*, July 10.

Thomas, Cal. 2005. Hurricane blows away dust on long history of corruption. *Los Angeles Times Syndicate,* reprinted in the *Arizona Republic*, Sept. 9.

Thomas, Evan et al. 2005. The lost city. *Newsweek,* Sept. 12.

Thompson, Mark 2005. The director. *Time,* Sept. 19.

Vaillant, George, Ed. 1986. *Empirical Studies of Ego Mechanisms and Defense.* Washington, D.C.: American Psychiatric Assoc. Press.

Wolffe, Richard. 2005. Yet another gulf war. *Newsweek,* Sept. 12.

Yochelson, Samuel and Stanton Samenow. 1976. *Criminal Personality* (3 vol.). New York: Jason Aronson.

Appendix A

Community Disaster Assessment

◆

Community Vulnerability Assessment

The Community Vulnerability Assessment Tool developed in North Carolina is an informational aid designed to assist communities in their efforts to reduce hazard vulnerability through strategies relating to awareness, education, and mitigation. This product contains a methodology that helps State and local governments determine and prioritize their locality's vulnerabilities to coastal hazards.

Physical factors such as the location of critical facilities and infrastructure relative to high-risk areas, the distribution of vulnerable populations such as the elderly, poor and under-insured, significant environmental resources and the vulnerability of primary economic sectors are all included as issues for consideration.

This tool is made available on the Internet through the NOAA (National Oceanic and Atmospheric Administration housed with the U.S. Department of Commerce).

Step One: Note hazards in your community and prioritize them.

Step Two: Assign some kind of scores within risk consideration areas according to the seriousness or frequency of the hazard.

Step Three: Identify critical facilities by category. This analysis focuses on determining the vulnerability of key individual facilities or resources within the community. Conduct an analysis for every structure in a community that you consider critical. Critical facility categories:

- Shelters
- Schools
- Hospitals and Nursing Homes
- Fire and Rescue
- Police
- Utilities
- Communications
- Transportation
- Government

Next, collect information about them:

- Facility type
- Facility name
- Street address
- City
- State
- Zip
- Owner/operator
- Contact name
- Contact title
- Contact telephone
- 24-hour telephone

- Township
- Fire district

Then identify intersections of critical facilities with high-risk areas.

To help prioritize potential hazard mitigation for critical facilities, you should identify risk area scores for each of the critical facilities. Overlay the critical facilities with the hazard risk areas. For each critical facility, conduct an individual assessment addressing the location of the facility relative to the hazard risk areas and the potential vulnerability of the facility to the impacts of each hazard. The facility assessment should be designed with questions that address the elements of vulnerability you are most concerned about, and the results should be weighted accordingly (i.e., structural vulnerability vs. operational vulnerability). If you are focusing on minimizing repetitive losses, you may wish to investigate damage history.

Step Four: Identify areas of special consideration.

Special consideration areas are those locations (preferably at the neighborhood level) where individual resources are minimal and personal resources for dealing with hazards can be extremely limited. These areas could be most dependent on public resources after a disaster and thus could be good investment areas for hazard mitigation activities. Identify special consideration areas by utilizing existing low-to-moderate income designations for community development grants or by analyzing key census data categories.

Examples:

- Minority Populations:
- Percent Households below Poverty.
- Percent Population over Age 65
- Percent Single Parent with Children

- Percent No High School Diploma
- Percent Households with Public Assistance Income
- Percent Housing Rental
- Percent Housing Units with No Vehicle Available

Step Five: Identify economic sectors.

The purpose of this analysis is to identify your economic vulnerability to hazard impacts. This first step focuses on identifying the major sectors of your economy and mapping primary centers of activity in those sectors. These economic centers are areas where hazard risks could have major impacts on your local economy and therefore would be ideal locations for targeting certain hazard mitigation strategies. To further target areas for potential hazard mitigation activities, you should identify economic centers that are located in high-risk areas. Then overlay the economic center maps with the hazard risk areas. The largest employers should also be listed along with their vulnerabilities and contact information.

Step Six: Identify secondary hazard risk consideration sites and key environmental resource sites.

The purpose of this analysis is to identify locations where there is potential for secondary environmental impacts from natural hazards and to target vulnerable locations for hazard mitigation activities. Secondary impacts occur when natural hazard events create new hazards such as toxic releases or hazardous spills. Identify key sites in your community where hazardous or toxic materials exist. To further target areas for potential hazard mitigation activities, you should identify secondary risk sites that are located in high natural hazard risk areas. Overlay the environmental sites with the hazard risk areas.

Step Seven: Identify areas of undeveloped land and their intersection with high-risk areas.

The purpose of this analysis is to identify opportunities beyond the existing built environment for reducing future hazard vulnerability. Identify the large tracts of undeveloped land in your community and, if possible, any future plans for growth. Overlay this information with the risk areas. Identify the type of land cover, land cover change over time (if possible), and zoning for all undeveloped land in high-hazard areas.

This information should provide an overview of the potential for future development in high-risk area locations. With this information, you can develop mitigation strategies that specifically target new development.

It is possible to obtain aggregate data on National Flood Insurance Program policies from the Federal Emergency Management Agency (FEMA). By establishing a ratio of policy-holders to households you can identify areas where education and outreach activities may be targeted as mitigation strategies.

Appendix B

Personal Disaster Assessment

✦

Guidelines for Hurricane Preparations

Note: The following information in Appendices A and B have been taken from information distributed to the public by NOAA, FEMA, Red Cross, EPA, and other agencies who give advice about how to handle emergencies.

To prepare for a hurricane, you should take the following measures:

- Make plans to secure your property. Permanent storm shutters offer the best protection for windows. A second option is to board up windows with 5/8" marine plywood, cut to fit and ready to install. Tape does not prevent windows from breaking.

- Install straps or additional clips to fasten securely your roof to the frame structure. This will reduce roof damage.

- Be sure trees and shrubs around your home are well trimmed.

- Clear loose and clogged rain gutters and downspouts.

- Determine how and where to secure your boat.

- Consider building a safe room.

If, and when, a hurricane is in your area, you should:

- Listen to the radio or TV for information.

- Secure your home, close storm shutters, and secure outdoor objects or bring them indoors.

- Turn off utilities if instructed to do so. Otherwise, turn the refrigerator thermostat to its coldest setting and keep its doors closed.

- Turn off propane tanks.· Avoid using the phone, except for serious emergencies.

- Moor your boat if time permits.

- Ensure a supply of water for sanitary purposes such as cleaning and flushing toilets. Fill the bathtub and other large containers with water.

You should evacuate under the following conditions:

- If you are directed by local authorities to do so. Be sure to follow their instructions.

- If you live in a mobile home or temporary structure—such shelters are particularly hazardous during hurricanes no matter how well fastened to the ground.

- If you live in a high-rise building—hurricane winds are stronger at higher elevations.

- If you live on the coast, on a floodplain, near a river, or on an inland waterway.

- If you feel you are in danger.

If you are unable to evacuate, go to your wind-safe room. If you do not have one, follow these guidelines:

- Stay indoors during the hurricane and away from windows and glass doors.

- Close all interior doors—secure and brace external doors.

- Keep curtains and blinds closed. Do not be fooled if there is a lull; it could be the eye of the storm—winds will pick up again.

- Take refuge in a small interior room, closet, or hallway on the lowest level.

- Lie on the floor under a table or another sturdy object.

Hurricanes are classified into five categories based on their wind speed, central pressure, and damage potential.

Category 1 has winds of 74-95 mph with 4-5 foot storm surges.

Category 2 has winds of 96-110 mph with 6-8 foot storm surges.

Category 3 has winds of 111-130 mph with 9-12 foot storm surges.

Category 4 has winds of 131-155 mph with 13-18 foot storm surges.

Category 5 has winds of 155 mph plus with 18 foot plus storm surges.

Since 1953, the following states have had direct hits by hurricanes:

60 Florida
38 Texas
28 North Carolina
27 Louisiana
14 South Carolina
12 Alabama
9 Mississippi, Maine, New York
5 Rhode Island
4 Virginia

HURRICANE WARNINGS

- WATCH: Hurricane conditions are *possible* in the specified area of the WATCH, usually within 36 hours.

- WARNING: Hurricane conditions are *expected* in the specified area of the WARNING, usually within 24 hours.

HURRICANE PREPARATIONS FOR YOUR HOME

- Locate a safe room or the safest areas in your home for each hurricane hazard. In certain circumstances, the safest areas may not be your home but within your community.

- Determine escape routes from your home and places to meet. These should be measured in tens of miles rather than hundreds of miles.

- Have an out-of-state friend as a family contact, so all your family members have a single point of contact.

- Make a plan now for what to do with your pets if you need to evacuate.

- Post emergency telephone numbers by your phones and make sure your children know how and when to call 911.

- Check your insurance coverage—flood damage is not usually covered by homeowners insurance.

- Stock non-perishable emergency supplies and a Disaster Supply Kit.

- Use a NOAA weather radio. Remember to replace its battery every 6 months, as you do with your smoke detectors.

- Take First Aid, CPR and disaster preparedness classes.

DISASTER SUPPLY KIT

- First aid kit and essential prescription medications.

- Canned food and can opener.

- At least three gallons of water per person.

- Protective clothing, rainwear, and bedding or sleeping bags.

- Battery-powered radio, flashlight, and extra batteries.

- Special items for infants, elderly, or disabled family members.

- Written instructions on how to turn off electricity, gas and water if authorities advise you to do so.

- Food for 3 to 7 days

- Cooking tools/fuel/paper plates/utensils

- Blankets/pillows

- Clothing/rain gear/sturdy shoes

- Toiletries/hygiene items/moisture wipes

- Radio-Battery operated and NOAA weather radio

- Cash/ATM cards

- Keys

- Toys/games/books

- Important documents in waterproof container or plastic bag: insurance, medical records, social security card, etc.

- Tools/knives/scissors

- Keep vehicle tanks filled

- Pet care items: immunization, medication, food, water, leash, muzzle

Obviously, if you are ordered to evacuate, you cannot take all of the things in these lists with you. You will have to take only the most important things, things that you can carry. But it is wise to have all of these things located in one part of your house so that they can quickly be gotten and are there if you are trapped in your house for some time, without other help available. Here are some more things to consider:

Aspirin or non-aspirin pain reliever
Anti-diarrhea medication
Antacid (for stomach upset)
Syrup of Ipecac (used to induce vomiting if needed)
Activated charcoal (used to neutralize poisons)
Tools and Supplies
Mess kits, or paper cups, plates, and plastic utensils
Emergency preparedness manual
Battery-operated radio and extra batteries
Flashlight and extra batteries
Cash or traveler's checks, change
Non-electric can opener, utility knife
Fire extinguisher: small canister ABC type
Tube tent
Pliers
Tape
Compass
Matches in a waterproof container
Aluminum foil

HOW TO STORE EMERGENCY WATER SUPPLIES

You can store your water in thoroughly washed plastic, glass, fiberglass, or enamel-lined metal containers. Never use a container that has held toxic substances, because tiny amounts may remain in the container's pores. Sound plastic containers, such as soft drink bottles, are best. You can also purchase food-grade plastic buckets or drums.

Before storing your water, treat it with a preservative, such as chlorine bleach, to prevent the growth of microorganisms. Use liquid bleach that contains 5.25 percent sodium hypochlorite and no soap. Add four drops of bleach per quart of water (or two scant teaspoons per 10 gallons), and stir. Seal your water containers tightly, label them and store them in a cool, dark place.

EMERGENCY OUTDOOR WATER SOURCES

If you need to seek water outside your home, you can use these sources. But purify the water before drinking it.

- Rainwater
- Streams, rivers and other moving bodies of water
- Ponds and lakes
- Natural springs

Avoid water with floating material, an odor, or dark color. Saltwater can be used only if you distill it first which is described below.

THREE WAYS TO PURIFY WATER

In addition to having a bad odor and taste, contaminated water can contain microorganisms that cause diseases such as dysentery, cholera, typhoid, and hepatitis. You should therefore purify all water of uncertain purity before using it for drinking, food preparation, or hygiene.

There are many ways to purify water. None are perfect. Often the best solution is a combination of methods. Before purifying, let any suspended particles settle to the bottom, or strain them through layers of paper towel or clean cloth.

Three easy purification methods are outlined below. These measures will kill microbes but will not remove other contaminants such as heavy metals, salts, most other chemicals, and radioactive fallout.

Boiling is the safest method of purifying water. Bring water to a rolling boil for 10 minutes, keeping in mind that some water will evaporate. Let the water cool before drinking.

Boiled water will taste better if you put oxygen back into it pouring it back and forth between two containers. This will also improve the taste of stored water.

Chlorination uses liquid chlorine bleach to kill microorganisms. Add two drops of bleach per quart of water (four drops if the water is cloudy), stir, and let stand for 30 minutes. If the water does not taste and smell of chlorine at that point, add another dose and let stand another 15 minutes.

If you do not have a dropper, use a spoon and a square-ended strip of paper or thin cloth about 1/4 inch by 2 inches. Put the strip in the spoon with an end hanging down about 1/2 inch below the scoop of the spoon. Place bleach in the spoon and carefully tip it. Drops the size of those from a medicine dropper will drip off the end of the strip.

Purification tablets release chlorine or iodine. They are inexpensive and available at most sporting goods stores and some drugstores. Follow the package directions. Usually one tablet is enough for one quart of water. Double the dose for cloudy water.

MORE RIGOROUS WATER PURIFICATION METHODS

While the three methods described above will remove only microbes from water, the following two purification methods will remove other contaminants. Distillation will remove microbes, heavy metals, salts, and most other chemicals.

Distillation involves boiling water and then collecting the vapor that condenses back to water. The condensed vapor will not include salt and other impurities. To distill, fill a pot halfway with water. Tie a cup to the handle on the pot's lid so that the cup will hang right-side-up when the lid is upside-down (make sure the cup is not dangling into the water) and boil the water for 20 minutes. The water that drips from the lid into the cup is distilled.

If activity is reduced, healthy people can survive on half their usual food intake for an extended period and without any food for many days. Food, unlike water, may be rationed safely, except for children and pregnant women.

If your water supply is limited, try to avoid foods that are high in fat and protein, and don't stock salty foods, since they will make you thirsty. Try to eat salt-free crackers, whole grain cereals, and canned foods with high liquid content.

You don't need to go out and buy unfamiliar foods to prepare an emergency food supply. You can use the canned foods, dry mixes and other staples on your cupboard shelves. In fact, familiar foods are important. They can lift morale and give a feeling of security in time of stress. Also, canned foods won't require cooking, water or special preparation.

FIRST AID KIT

Assemble a first aid kit for your home and one for each car.

(20) adhesive bandages, various sizes.
(1) 5" x 9" sterile dressing.
(1) conforming roller gauze bandage.
(2) triangular bandages.
(2) 3 x 3 sterile gauze pads.

(2) 4 x 4 sterile gauze pads.

(1) roll 3" cohesive bandage.

(2) germicidal hand wipes or waterless alcohol-based hand sanitizer.

(6) antiseptic wipes.

(2) pair large medical grade non-latex gloves.

Adhesive tape, 2" width.

Anti-bacterial ointment.

Cold pack.

Scissors (small, personal).

Tweezers

CPR breathing barrier such as a face shield.

PERSONAL EVACUATION PLAN

- Identify ahead of time where you could go if you are told to evacuate. Choose several places—a friend's home in another town, a motel, or a shelter.

- Keep handy the telephone numbers of these places as well as a road map of your locality. You may need to take alternative or unfamiliar routes if major roads are closed or clogged.

- Prepare to bring inside any lawn furniture, outdoor decorations or ornaments, trash cans, hanging plants, and anything else that can be picked up by the wind.

- Prepare to cover all windows of your home. If shutters have not been installed, use precut plywood as described above. *Note:* Tape does not prevent windows from breaking, so taping windows is not recommended.

- Fill your car's gas tank.

PREPARE YOUR HOME FOR A HURRICANE

The most important precaution you can take to reduce damage to your home and property is to protect the areas where wind can enter. According to recent wind technology research, it's important to strengthen the exterior of your house so wind and debris do not tear large openings in it. You can do this by protecting and reinforcing these five critical areas: roof, straps, shutters, doors, and garage doors. A great time to start securing or retrofitting your house is when you are making other improvements or adding an addition. Remember that building codes reflect the lessons experts have learned from past catastrophes. Contact the local building code official to find out what requirements are necessary for your home improvement projects.

- Install hurricane shutters or purchase precut ½" outdoor plywood boards for each window of your home. Install anchors for the plywood and pre-drill holes in the plywood so that you can put it up quickly.

- Make trees more wind resistant by removing diseased and damaged limbs, then strategically removing branches so that wind can blow through.

CHECK YOUR INSURANCE

Flood damage is not usually covered by homeowners insurance. Do not make assumptions. Check your policy. The National Flood Insurance Program is a pre-disaster flood mitigation and insurance protection program designed to reduce the escalating cost of disasters. The National Flood Insurance Program makes federally backed flood insurance available to residents and business owners. Call them at 1-888-call-flood, extension 445.

About the Author

Diane Holloway, Ph.D., Dallas psychologist, coordinated the first Dallas police assessment center, was appointed the first Dallas "Drug Czar" by the Mayor, and served as consultant and trainer to fire, police, and city governments across the country. Earlier she also served as a management consultant and trainer on personnel selection with some of America's largest corporations.

She wrote *The Mind of Oswald; Dallas and the Jack Ruby Trial; American History in Song;* and *Analyzing Leaders, Presidents and Terrorists.* She co-authored *Before You Say 'I Quit'; Who Killed Detroit; Jacuzzi: A Father's Invention to Ease a Son's Pain; Growing Up a Shadow: Abuse Recovery;* and *I Did Not Burn the Church Down…I Only Started the Fire.* She served on the local fire district board and highly values emergency services.

Johannes F. Spreen, B.S., M.P.A., and Ph.D. (ABD) was in law enforcement from 1941-1985, starting as a New York City policeman, and rose to Inspector and Command of Operations. He was Detroit Police Commissioner; Sheriff of Oakland County, Michigan; newspaper columnist, and a professor at John Jay College and Mercy College. He has written *American Law Enforcement Does Not Serve or Protect, American Police Dilemma: Protectors or Enforcers* and co-authored *Who Killed Detroit? Other Cities Beware.*

Bob Cheney taught history for 38 years in Dallas public schools, and the Dallas County Community Colleges. He is retired and living in Arizona, but still teaches locally and for Elderhostel. He has written *Interrupted Lives: Hood's Texas Brigade; Tragedy in Black and White; Brushes with Greatness: A Chronicle of Five Generations of American Life;* and *Stampin' Out Ignorance.*

Index

A

Accardo, Paul 51, 104
Adams, Lee 30
Allen, Thad 77
Avery, Crissy 16

B

Bailey, Chris 115
Barry, John 28, 68
Bin Laden, Osama 3, 12, 48
Biscoe, Robert vii
Blanco, Kathleen 8, 74, 121
Brown, Michael 5, 18, 74
Bush, George W. 3, 49, 67, 89

C

Carlson, Peter 68
Castro, Fidel 11, 12
Celestine, Lawrence 52
Clarke, Ronald 100
Cleaver, Eldridge 112
Clemens, Samuel 24, 32
Clinton, Bill 49
Clinton, Hillary Rodham 9
Cohen, Lawrence 100
Compass, Eddie 52, 82
Coolidge, Calvin 28
Cornish, Derek 100

D

Dean, Howard 3
Defillo, Marlon 51
Diamond, Jared 128

DiLeonardo, Robert 90
Douglas, John 100, 101
DuBois, Blanche 103
Duke, David 68
Dulles, John Foster 50

E

Edwards, Edwin 68, 91

F

Farrakhan, Louis 89
Farrior, Don vii
Faulkner, William 32
Felson, Marcus 100
Ferber, Edna 31
Fischetti, Mark 36
Fitzgerald, Edward 98
Foti, Charles 64
Franklin, Benjamin 26
Freud, Sigmund 1, 99, 151

G

Gore, Al 5
Gusman, Marlin 64

H

Hamill, Pete 80
Harte, Bret 114
Hastert, Dennis 115
Hitler, Adolf 3, 12
Hoover, Herbert 28

J

Jackson, Jesse 81, 89

Jacuzzi, Ken vii
Jefferson, Blind Lemon 29
Jefferson, Thomas 20, 26

K
Kelling, George 102
Kennedy, John 4, 12
Kern, Jerome 31
Kohlberg, Lawrence 99

L
Landrieu, Mary 72, 78
Lee, Robert E. 27, 28
Lewis, Meriwether 27
Lincoln, Abraham 140
Lindbergh, Charles 31
Lippmann, Walter 67
Loevinger, Jane 99
London, Jack 25
Lucretius 93

M
McCain, John 72, 121
McPhee, John 34
Melville, Herman 117
Mill, John Stuart 98

N
Nagin, Ray 7, 14, 53, 61, 68, 72, 121, 137
Napoleon 26
Napolitano, Janet 108
Newsom, Gavin 109
Noah 21

P
Parent, Charles 55
Patton, Charlie 30
Penney, Steve vii

Perry, Rick 122
Plato 97, 98

R
Revkin, Andrew 117
Rice, Condoleezza 5
Roland, Walter 29

S
Samenow, Stanton 101, 152
Samuelson, Robert J. 92
Santayana, George 125
Sawyer, Diane 48, 97
Schwartz, John 117
Scutari, Chip 108
Shakespeare, William 33, 98
Socrates 97, 98, 99
Stillwell, Valentina 15
Strain, Jack 65
Strouse, Charles 30
Suhayda, Joe 35

T
Thomas, Lyda Ann 123
Thompson, Mark 13
Trittin, Jurgen 40
Truman, Harry 12, 18, 45, 140
Tully, Edward J. 57
Twain, Mark 24, 151

W
Wald, Matthew 117
Waldron, Patrick 116
White, John 78
Williams, Tennessee xiii, 103
Wilson, James Q. 102

Y
Yochelson, Samuel 101

978-0-595-37391-8
0-595-37391-7

www.ingramcontent.com/pod-product-compliance
Lightning Source LLC
Chambersburg PA
CBHW020413290526
45785CB00002B/546